Statistics Workbook

2nd Edition with Online Practice

by Deborah J. Rumsey, PhD

Statistics Workbook For Dummies®, 2nd Edition with Online Practice

Published by: **John Wiley & Sons, Inc.**, 111 River Street, Hoboken, NJ 07030-5774, www.wiley.com

Library of Congress Control Number: 2019931550

ISBN 978-1-119-54751-8 (pbk); ISBN 978-1-119-54767-9 (ebk); ISBN 978-1-119-54768-6

Manufactured in the United States of America

C10013332_082619

Contents at a Glance

Table of Contents

Introduction

Perhaps you're taking a statistics class, or you're about to take one. You may understand some of the basic ideas, but you have questions and want a place to go for a little extra help to give you an edge. And you also want a heads-up as to what instructors really think about when they write their exams. Well, look no further; help has arrived in the form of *Statistics Workbook For Dummies*, 2nd Edition.

This workbook helps you become more comfortable with and confident about statistics. Through plenty of practical problems that take you from step one all the way to your final exam, you review the concepts you know, identify areas where you need to focus more work, and address the little things that can make the difference between a B and an A.

As a statistics professor who has taught tens of thousands of students over the years, I have noticed that certain problems keep cropping up and causing my pen to take points off exams over and over again. And believe me: I want nothing more than to put my red pen away. So I give you all my secrets about what professors really want you to know, the kinds of questions they ask, and the types of answers they love and hate to see (so you can avoid the latter). And I focus only on the topics that you absolutely need to know, with minimal background information.

About This Book

The major objectives of this workbook are for you to understand, calculate, and interpret the most common statistical formulas and techniques; get a handle on basic probability; gain confidence with difficult statistical topics such as the central limit theorem and *p*-values; know which statistical technique to use in different situations (for example, when to employ what kind of confidence interval); and evaluate and pinpoint problems with studies, polls, and experiments.

Although I wrote this workbook to serve as a companion to *Statistics for Dummies*, 2nd Edition (also published by Wiley and written by yours truly), this workbook works quite well with any introductory statistics textbook.

You may be asking how this workbook is different from other workbooks on the shelf. Well, here are a few ways, listed in order of importance:

- **Plenty of excellent practice problems** to lead you down the path of examination success, chosen by me, a card-carrying member of the "million statistics exam question writing and grading" club. I provide all answers at the end of each chapter.

- **Workspace** for you to work through the problems directly in the section you're working on, so you can easily refer to your notes later when you need them.
- **Not only answers, but also clear, complete explanations to go with them.** Explanations help you know exactly how to approach a problem, what information you need to solve it, and common problems you need to avoid.
- **A view inside a professor's mind** to help you determine the most popular questions, the answers we look for, and the answers that make us pull our hair out.
- **Tips, strategies, and warnings** based on my vast experience with students of all backgrounds and learning styles (and my grading experience).
- **An example accompanying each section,** directly followed by the solution. Use the example as a reference when you work the other problems.
- **A focus on problem-solving skills** to help you develop a problem-solving strategy when you take exams. I don't show you how I would do the problems; I help you see how you can do the problems. And believe me, there's a big difference!
- **The nonlinear approach** allows you to skip around in the workbook and still have easy access to and understanding of any given topic.
- **Understandable language** to help you process, remember, and put into practice statistical definitions, techniques, and processes.
- **Clear and concise step-by-step procedures** that intuitively explain how to work through statistics problems and remember the process.

I also used a few conventions while writing this book that you should be aware of:

- The most important convention that you need to be aware of deals with my dual use of the word "statistics." In some situations, I refer to statistics as a subject of study or as a field of research. For example, "Statistics is really quite an interesting subject!" (Note I said statistics "is" in this case.) In other situations, I refer to statistics as the plural of statistic, in a numerical sense. For example, "The most common statistics are the mean and the standard deviation." (Notice my use of the word "are" in this case.)
- I also use data in a plural form ("the data are" rather than "the data is"). The battle rages on between statisticians over which way is right, but I go with the plural form.
- I use Ho to represent the null hypothesis in a hypothesis test. Although this is a commonly used notation, others might use the notation H_0 to mean the same thing.
- I use * to indicate a multiplication sign.

Foolish Assumptions

This book is for you if you have some exposure to statistics already and want more opportunities to enjoy success through additional practice of the skills and techniques. Or perhaps you're taking a statistics class and could use some extra support (and insider information). Or maybe you just really want to understand *p*-values because they keep you awake at night (been there, done that).

Note: If you're totally new to the subject of statistics, I suggest that you first read *Statistics for Dummies*, 2nd Edition, (Wiley), because I cover the various concepts of statistics in much more detail in that book (but any introductory text will suffice). After you feel comfortable and confident with the material, you can try the problems in this workbook. Or, as an alternative, you can use this workbook to practice along with what you read in *Statistics For Dummies*, 2nd Edition.

Icons Used in This Book

Icons in this workbook draw your attention to certain features that occur on a regular basis. Think of them as road signs that you encounter on a trip. Here are the road signs you encounter on your journey through this workbook.

Each section of this workbook begins with a brief overview of the topic. After the intro, you see an example problem with a fully worked solution for use as a reference as you work the practice problems. You can quickly locate the example problems by looking for this icon.

I use this icon for particular ideas that I hope you'll remember long after you read this workbook.

This icon points out helpful hints, ideas, or shortcuts that save you time or give you alternative ways to think about a particular concept. I also use this icon to "get down to the nitty-gritty" discussing the types of questions your instructor may ask you and why, revealing what instructors really look for in your answers, and giving you a heads-up on the types of errors that really make them nuts (so you can avoid them at all costs).

This icon refers to specific ways that you may get tripped up while working a certain kind of problem and how to avoid those problems. Commit these items to memory while it still doesn't cost you any points (in other words, before the exam takes place).

Beyond the Book

Be sure to check out the free Cheat Sheet for a handy guide that covers tips and tricks for answering statistics questions. To get this Cheat Sheet, simply go to `www.dummies.com` and enter "Statistics Workbook For Dummies" in the Search box.

You also have the opportunity to complete online quizzes for Chapters 1 through 18 that test your knowledge of the concepts in each chapter. To gain access to the online practice, all you have to do is register by following these simple steps:

1. **Find your PIN access code located on the inside front cover of this book.**
2. **Go to `www.dummies.com` and click** Activate Now.
3. **Find your product (*Statistics Workbook For Dummies*, 2nd Edition) and then follow the on-screen prompts to activate your PIN.**

Now you're ready to go! You can go back to the program at testbanks.wiley.com as often as you want — simply log on with the username and password you created during your initial login. No need to enter the access code a second time.

Tip: If you have trouble with your PIN or can't find it, contact Wiley Product Technical Support at 877-762-2974 or go to support.wiley.com.

Where to Go from Here

I wrote this workbook in a nonlinear way, so you can start anywhere and still understand what's happening. However, I can make some recommendations to readers who are interested in knowing where to start:

- If you want to get right into the number-crunching aspects of statistics (finding the mean, median, standard deviation, and so on), I suggest starting with Part 1.
- If you want to break down the normal distribution or the central limit theorem, go to Part 2.
- If you're most worried about confidence intervals and hypothesis tests, jump to Part 3.
- If you want to develop your skills evaluating and making sense of the results of medical studies, polls, surveys, and experiments, start with Part 4.
- If you want to nail down data collected on two variables (correlation and the like), head directly to Part 4.
- If you want tips on math, common statistical formulas, or ways to spot statistical mistakes, head to Part 5.

1 Getting Off to a Statistically Significant Start

IN THIS PART . . .

Get down to the basics of number crunching.

Make and interpret charts and graphs.

Crank out and understand descriptive statistics.

Develop important skills for critiquing others' statistics.

» Highlighting the difference between frequencies and relative frequencies

» Interpreting and evaluating tables

Chapter 1 Summarizing Categorical Data: Counts and Percents

Categorical data is data in which individuals are placed into groups or categories — for example gender, region, or type of movie. Summarizing categorical data involves boiling down all the information into just a few numbers that tell its basic story. Because categorical data involves pieces of data that belong in categories, you have to look at how many individuals fall into each group and summarize the numbers appropriately. In this chapter, you practice making, interpreting, and evaluating frequency and relative frequency tables for categorical data.

Counting On the Frequency

One way to summarize categorical data is to simply count, or tally up, the number of individuals that fall into each category. The number of individuals in any given category is called the frequency (or count) for that category. If you list all the possible categories along with the frequency for each, you create a frequency table. The total of all the frequencies should equal the size of the sample (because you place each individual in one category).

See the following for an example of summarizing data by using a frequency table.

EXAMPLE

Q. Suppose that you take a sample of 10 people and ask them all whether they own a cellphone. Each person falls into one of two categories: yes or no. The data are shown in the following table.

Person #	Cellphone	Person #	Cellphone
1	Y	6	Y
2	N	7	Y
3	Y	8	Y
4	N	9	N
5	Y	10	Y

a. Summarize this data in a frequency table.

b. What's an advantage of summarizing categorical data?

A. Data summaries boil down the data quickly and clearly.

a. The frequency table for this data is shown in the following table.

b. A data summary allows you to see patterns in the data, which aren't clear if you look only at the original data.

Own a Cellphone?	Frequency
Y	7
N	3
Total	10

1 You survey 20 shoppers to see what type of soft drink they like best, Brand A or Brand B. The results are: A, A, B, B, B, B, B, B, A, A, A, B, A, A, A, A, B, B, A, A. Which brand do the shoppers prefer? Make a frequency table and explain your answer.

2 A local city government asks voters to vote on a tax levy for the local school district. A total of 18,726 citizens vote on the issue. The yes count comes in at 10,479, and the rest of the voters said no.

a. Show the results in a frequency table.

b. Why is it important to include the total number at the bottom of a frequency table?

3 A zoo asks 1,000 people whether they've been to the zoo in the last year. The surveyors count that 592 say yes, 198 say no, and 210 don't respond.

a. Show the results in a frequency table.

b. Explain why you need to include the people who don't respond.

4 Suppose that instead of showing the number in each group, you show just the percentage (called a *relative frequency*). What's one advantage a relative frequency table has over a frequency table?

Relating with Percentages

Another way to summarize categorical data is to show the percentage of individuals who fall into each category, thereby creating a relative frequency. The *relative frequency* of a given category is the frequency (number of individuals in that category) divided by the total sample size, multiplied by 100 to get the percentage. For example, if you survey 50 people and 10 are in favor of a certain issue, the relative frequency of the "in-favor" category is $10 \div 50 = 0.20$ times 100, which gives you 20 percent. If you list all the possible categories along with their relative frequencies, you create a *relative frequency table*. The total of all the relative frequencies should equal 100 percent (subject to possible round-off error).

See the following for an example of summarizing data by using a relative frequency table.

EXAMPLE

Q. Using the cellphone data from the following table, make a relative frequency table and interpret the results.

Person #	Cellphone	Person #	Cellphone
1	Y	6	Y
2	N	7	Y
3	Y	8	Y
4	N	9	N
5	Y	10	Y

A. The following table shows a relative frequency table for the cellphone data. Seventy percent of the people sampled reported owning cellphones, and 30 percent admitted to being technologically behind the times.

Own a Cellphone?	Relative Frequency
Y	70%
N	30%

You get the 70 percent by taking $7 \div 10 * 100$, and you calculate the 30 percent by taking $3 \div 10 * 100$.

5 You survey 20 shoppers to see what type of soft drink they like best, Brand A or Brand B. The results are: A, A, B, B, B, B, B, B, A, A, A, B, A, A, A, A, B, B, A, A. Which brand do the shoppers prefer?

a. Use a relative frequency table to determine the preferred brand.

b. In general, if you had to choose, which is easier to interpret: frequencies or relative frequencies? Explain.

6 A local city government asked voters in the last election to vote on a tax levy for the local school district. A record 18,726 voted on the issue. The yes count came in at 10,479, and the rest of the voters checked the no box. Show the results in a relative frequency table.

A zoo surveys 1,000 people to find out whether they've been to the zoo in the last year. The surveyors count that 592 say yes, 198 say no, and 210 don't respond. Make a relative frequency table and use it to find the *response rate* (percentage of people who respond to the survey).

Name one disadvantage that comes with creating a relative frequency table compared to using a frequency table.

Interpreting Counts and Percents with Caution

Not all summaries of categorical data are fair and accurate. Knowing what to look for can help you keep your eyes open for misleading and incomplete information.

Instructors often ask you to "interpret the results." In this case, your instructor wants you to use the statistics available to talk about how they relate to the given situation. In other words, what do the results mean to the person who collects the data?

REMEMBER

With relative frequency tables, don't forget to check whether all categories sum to 1 or 100 percent (subject to round-off error), and remember to look for some indicator as to total sample size.

See the following for an example of critiquing a data summary.

EXAMPLE

Q. You watch a commercial where the manufacturer of a new cold medicine ("Nocold") compares it to the leading brand. The results are shown in the following table.

How Nocold Compares	Percentage
Much better	47%
At least as good	18%

a. What kind of table is this?

b. Interpret the results. (Did the new cold medicine beat out the leading brand?)

c. What important details are missing from this table?

A. Much like the cold medicines I always take, the table about "Nocold" does "Nogood."

a. This table is an incomplete relative frequency table. The remaining category is "not as good" for the Nocold brand, and the advertiser doesn't show it. But you can do the math and see that $100\% - (47\% + 18\%) = 35\%$ of the people say that the leading brand is better.

b. If you put the two groups together, 65% of the patients say that Nocold is at least as good as the leading brand, and almost half of the patients say Nocold is much better.

c. What's missing? The remaining percentage (to keep all possible results in perspective). But more importantly, the total sample size is missing. You don't know whether the surveyors sampled 10 people, 100 people, or 1,000 people. This means that the precision of the results is unknown. (Precision means how consistent the results will be from sample to sample; it's related to sample size, as you see in Chapter 10.)

9 Suppose that you ask 1,000 people to identify from a list of five vacation spots which ones they've already visited. The frequencies you receive are Disney World: 216; New Orleans: 312; Las Vegas: 418; New York City: 359; and Washington, D.C.: 188.

a. Explain why creating a traditional relative frequency table doesn't make sense here.

b. How can you summarize this data with percents in a way that makes sense?

10 If you have only a frequency table, can you find the corresponding relative frequency table? Conversely, if you have only a relative frequency table, can you find the corresponding frequency table? Explain.

Answers to Problems in Summarizing Categorical Data

1. Eleven shoppers prefer Brand A, and nine shoppers prefer Brand B. The frequency table is shown in the following table. Brand A got more votes, but the results are pretty close.

Brand Preferred	Frequency
A	11
B	9
Total	20

2. Frequencies are fine for summarizing data as long as you keep the total number in perspective.

   a. The results are shown in the following table. Because the total is 18,726, and the yes count is 10,479, the no count is the difference between the two, which is $18,726 - 10,479 = 8,247$.

   b. The total is important because it helps keep the frequencies in perspective when you compare them to each other.

Vote	Frequency
Y	10,479
N	8,247
Total	18,726

3. This problem shows the importance of reporting not only the results of participants who respond but also what percentage of the total actually respond.

   a. The results are shown in the following table.

   b. If you don't show the nonrespondents, the total doesn't add up to 1,000 (the number surveyed). An alternative way to show the data is to base it on only the respondents, but the results would be biased. You can't definitively say that the nonrespondents would respond the same way as the respondents.

Gone to the Zoo in the Last Year?	Frequency
Y	592
N	198
Nonrespondents	210
Total	1,000

(4) Showing the percents rather than counts means making a relative frequency table rather than a frequency table. One advantage of a relative frequency table is that everything sums to 100 percent, making it easier to interpret the results, especially if you have a large number of categories.

(5) Relative frequencies do just what they say: They help you relate the results to each other (by finding percentages).

a. Eleven shoppers out of the 20 prefer Brand A, and nine shoppers out of the 20 prefer Brand B. The relative frequency table is shown in the following table. Brand A got more votes, but the results are pretty close, with 55 percent of the shoppers preferring Brand A, and 45 percent preferring Brand B.

Brand Preferred	Relative Frequency
A	55%
B	45%

b. You often have an easier time interpreting percents, because when you need to interpret counts, you have to put them in perspective in terms of "out of how many?"

(6) The results are shown in the following table. The yes percentage is $10{,}479 \div 18{,}726 = 55.96\%$. Because the total is 100%, the no percentage is $100\% - 55.96\% = 44.04\%$.

Vote	Relative Frequency
Y	55.96%
N	44.04%

(7) You can see the relative frequency table that follows this answer. Knowing the response rate is critical for interpreting the results of a survey. The higher the response rate, the better. The response rate is $59.2\% + 19.8\% = 79.0\%$ – the total percentage of people who responded in any way (yes or no) to the survey. (Note that 21% is the nonresponse rate.)

Gone to the Zoo in the Last Year?	Relative Frequency
Y	$592 \div 1{,}000 = 0.592 = 59.2\%$
N	19.8%
Nonrespondents	21.0%

(8) One disadvantage of a relative frequency table is that if you see only the percents, you don't know how many people participated in the study; therefore, you don't know how precise the results are. You can get around this problem by putting the total sample size somewhere at the top or bottom of your relative frequency table.

REMEMBER

When making a relative frequency table, include the total sample size somewhere on the table.

(9) Be careful about how you interpret tables where an individual can be in more than one category at the same time.

a. The frequencies don't sum to 1,000, because people have the option to choose multiple locations or none at all, so each person doesn't end up in exactly one group. If you take the grand total of all the frequencies (1,493) and divide each frequency by 1,493 to get a relative frequency, the relative frequencies sum to 1 (or 100 percent). But what does that mean? It makes it hard to interpret these percents because they don't account for the total number of people.

b. One way you can summarize this data is by showing the percentage of people who have been at each location separately (compared to the percentage who haven't been there before). These percents add up to 1 for each location. The following table shows the results summarized with this method. ***Note:*** The table isn't a relative frequency table; however, it uses relative frequencies.

Location	% Who Have Been There	% Who Haven't Been There
Disney World	$216 \div 1{,}000 = 21.6\%$	$100\% - 21.6\% = 78.4\%$
New Orleans	$312 \div 1{,}000 = 31.2\%$	68.8%
Las Vegas	$418 \div 1{,}000 = 41.8\%$	58.2%
New York City	$359 \div 1{,}000 = 35.9\%$	64.1%
Washington, D.C.	$188 \div 1{,}000 = 18.8\%$	81.2%

REMEMBER

Not all tables involving percents should sum to 1. Don't force tables to sum to 1 when they shouldn't; do make sure you understand whether each individual can fall under more than one category. In those cases, a typical relative frequency table isn't appropriate.

(10) You can always sum all the frequencies to get a total and then find each relative frequency by taking the frequency divided by the total. However, if you have only the percents, you can't go back and find the original counts unless you know the total number of individuals. Suppose that you know that 80 percent of the people in a survey like ice cream. How many people in the survey like ice cream? If the total number of respondents is 100, $80 = (100 * 0.80)$ people like ice cream. If the total is 50, you're looking at $40 = (50 * 0.80)$ positive answers. If the total is 5, you deal only with $4 = (5 * 0.80)$. This illustrates why relative frequency tables need to have the total sample size somewhere.

REMEMBER

Watch for total sample sizes when given a relative frequency table. Don't be misled by percentages alone, thinking they're always based on large sample sizes, because many are not.

» Interpreting your results properly

» Compiling the results for a data discussion

Chapter 2
Summarizing Quantitative Data: Means, Medians, and More

Before data are organized in a chart or graph, the first step is to summarize them — that is, find a few numbers and/or words that can tell the story of the data in a nutshell. For quantitative data, the most important characteristics are the shape of the data (which you see in Chapter 4), where the center is located, and how much variability or spread is in the data. You may also want to point out any *outliers* in the data (numbers that appear far from the rest). And like everything else in statistics, there's room for people to stretch the truth in how they choose to summarize their data (or in what they choose not to tell you). So it's good to know the big ideas of how data are summarized and what to look for in terms of interpreting and evaluating data summaries. That's what you practice in this chapter.

Finding and Interpreting Measures of Center

The most common way to summarize quantitative data is to describe where the center is. One way of thinking about what *center* means is to ask, "What's a typical value in this data set?" You can measure the center of a data set in different ways, and the method you choose can greatly influence the conclusions people make about the data.

The mean of a data set is also known as the *average*. To find the mean, add all the numbers in the data set and divide by the number of numbers. The notation for the sample mean is $\bar{x}$, and the formula for the sample mean is $\bar{x} = \frac{\sum_{i=1}^{n} x_i}{n}$. In this case the capital sigma stands for sum, and the subscript "i" starts at 1 and ends with n, so you are summing each value in the data set from x_1 to x_n. Then divide by n.

The *median* of a data set is the true middle value when the data are ordered from smallest to largest. To find the median, order the data and pick the middle number(s). If you have an odd number of numbers, only one value is in the middle. If you have an even number of numbers, you pinpoint two values in the middle and determine the average of the two to get the median.

See the following for an example of calculating the mean and median.

EXAMPLE

Q. Find the mean and the median of the following data set: 1, 6, 5, 7, 3, 2.5, 2, –2, 1, 0.

A. The mean is $1+6+5+7+3+2.5+2+-2+1+0=25.5$, divided by 10 (because you have 10 numbers), which equals 2.55. To find the median, order the numbers: –2, 0, 1, 1, 2, 2.5, 3, 5, 6, 7. Now find the middle number. In this, case there are two middle values: 2 and 2.5. Take the average: $(2+2.5)\div 2 = 4.5 \div 2 = 2.25$.

Does the mean have to be one of the numbers in the data set? Explain.

Does the median have to be one of the numbers in the data set? Explain.

3. Why do you have to order the data to calculate the median but not for the mean?

4. Suppose that you have an *outlier* in a data set (a number that stands out away from the rest). How does an outlier affect the mean and the median of that data set?

5. Suppose that you find the mean for a certain data set.

   a. Depending on what the data actually are, the mean should always lie between the largest and smallest values of the data set. Explain why.

   b. When can the mean be the largest value in the data set?

6. Give an example of two different data sets containing *three* numbers each that both have the same median and mean. Explain why the median isn't enough to tell the whole story about a data set.

7 Suppose that the mean and median salary at a company is \$50,000, and all employees get a \$1,000 raise.

a. What happens to the mean?

b. What happens to the median?

8 Suppose that the mean and median salary at a company is \$50,000, and all employees get a 10% raise.

a. What happens to the mean?

b. What happens to the median?

Finding and Interpreting Measures of Spread

Variation is one of the most important concepts in statistics. It measures how much the values in a sample or a population fluctuate. Values that appear close together indicate a small amount of variation. Values that are spread out indicate a large amount of variation.

A very crude measure of spread is the *range.* The statistical definition of range is the biggest number in the data set minus the smallest number. The range is a single value, not a pair of values, and is entirely based on only two numbers. Both numbers can be outliers, which is why range can be a crude measure of spread.

Because range is such a crude measure, by far the most common measure of variation is the *standard deviation.* The standard deviation represents the "typical" distance from any point in the data set to the mean. Roughly speaking, standard deviation gives you the average distance from the mean.

To find the standard deviation of data from a sample, you first find the mean (refer to the previous section). After that, follow these steps:

1. **Take each number in the data set, subtract the mean, and square the result.**
2. **Add up all these so-called "squared deviations" and divide by $n-1$ (n is the size of the data set).**
3. **Take the square root to undo the squaring you did earlier.**

The notation for sample standard deviation is s, and the formula is $s=\sqrt{\frac{\sum_{i=1}^{n}(x_i-\bar{x})^2}{n-1}}$.

See the following for an example of calculating and interpreting standard deviation.

EXAMPLE

Q. Find and interpret the standard deviation of the following data set: 1, 2, 3, 4, 5.

A. First, the mean of this data set is 3 (see the previous section in this chapter for mean info). After you calculate the mean, find the deviations from the mean and square them: $1-3=-2$, and -2 squared equals 4; $2-3=-1$, and -1 squared equals 1; $3-3=0$, and 0 squared equals 0; $4-3=1$, and 1 squared equals 1; and finally, $5-3=2$, and 2 squared equals 4. Sum these values up to get $4+1+0+1+4=10$. Divide 10 by $5-1$ (because $n=5$) to get $10\div4=2.5$. The final step is to take the square root of 2.5, which gives you $s=1.58$. This answer means the data are, on average, about 1.58 steps from the mean (3).

9 What's the smallest standard deviation you can figure, and when would that happen?

10 Choose four numbers from 1 to 5, with repetitions allowed, to create the largest standard deviation possible.

11 Suppose that the mean salary at a company is $50,000, and all employees get a $1,000 raise. What happens to the standard deviation?

12 Suppose that the mean salary at a company is $50,000, and all employees get a 10% raise. What happens to the standard deviation?

Using Percentiles and the Interquartile Range

When dealing with *skewed* data (data that aren't symmetric but rather lopsided off to one side), it's often better to work with the median as the measure of center, because it's not affected by the skewness as much as the mean is. And along this line, you can measure the spread of skewed data by focusing mainly on the range of the middle 50 percent of the data — called the *interquartile range*, or IQR. To understand IQR, you need to review percentiles — the *k*th percentile is a point in the data set where *k*% of the data lies below it. So if your height is at the 70th percentile, for example, that means 70 percent of the people are shorter than you. Certain numbers that represent the 25th and 75th percentiles have special names because they divide the data into quarters; they are called the first quartile, or Q1, and the third quartile, or Q3, respectively.

Note that computers can calculate quartiles for you, and that's how you'll normally get them done.

But, for example, suppose that you have 10 numbers (ordered) 1, 3, 5, 6, 7, 8, 10, 12, 13, 13. The Q3 would be the number that is three-fourths of the way through the data. One way to think of it is to find the median first (7.5 here), which would be the middle of the data. The median now divides the data into two halves, the upper half and the lower half. The Q1 would be the median of the lower half of the data (half of a half is a quarter), which here is the median of 1, 3, 5, 6, 7, which is 5. The Q3 would be the median of the upper half of the data; the median of 8, 10, 12, 13, 13 is 12. (Other computer programs may give slightly different answers to finding quartiles, but you get the big picture here.)

When you calculate the IQR you are ignoring any skewness or outliers that may lie on either end. To find the IQR, you find the range of the middle 50% of the data; you take Q3 – Q1. So if the 75th percentile (Q3) for height was 60 inches and the 25th percentile (Q1) was 53 inches, then the IQR for height would be $60 - 53 = 7$ inches.

EXAMPLE

Q. Find the first quartile, the third quartile, and the IQR of the numbers 1, 2, 3, 4, 5.

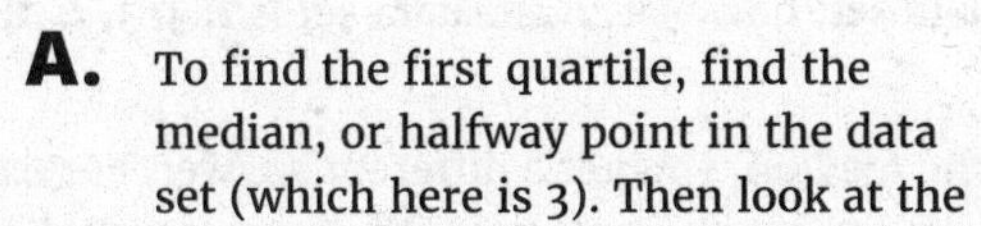

A. To find the first quartile, find the median, or halfway point in the data set (which here is 3). Then look at the lower half of the numbers and find the median of that, which here is 2, so the first quartile is 2. Similarly, the third quartile is the median of the upper half of the data, which here is 4. The IQR is $4 - 2 = 2$.

13 Find the median and the IQR of the numbers 1, 2, 3, 4, 5, 6.

14 Find the median and the IQR of the numbers 2, 2, 2, 2, 2.

15 Is the IQR affected by outliers or skewness? Why or why not?

Answers to Problems in Summarizing Quantitative Data

1. The mean (or average) doesn't have to be one of the numbers in the data set, but it can be. For example, in the data set 1, 2, the mean is 1.5, which isn't in the data set; however, in the data set 1, 2, 3, the mean is 2.

2. The median will be one of the numbers in the data set if the set has an odd number of values in it, because the set has one distinct middle value in that case. If the set has an even number of values, you find the median by averaging the two middle values, and the answer may or may not be one of the values in the data set. For example, if the data set is 1, 2, 3, 4, the median is 2.5, which isn't included in the data set; however, if the data set is 1, 3, 3, 4, the median is $(3+3) \div 2 = 3$, which is included.

3. If you don't order the data to find the median, you get a different answer. For example, look at the data set 1, 5, 2. The median is 2, but if you don't order the data, it would be 5. And if you reorder the same data set to be 2, 1, 5, you get a different answer for the median: 1. So you should always order the data from smallest to largest to always get the same answer for the median. For the mean, you add up all the values in the data set and divide by the size of the data set. Using the commutative property for addition (and you thought you'd never use algebra later in life!), you know that $a + b = b + a$. Even if you reorder the data, you still get the same sum. So you don't have to order the data to always get the same answer for the mean of a given data set.

4. Outliers attract the mean toward them and away from the rest of the data. For example, the mean and the median of the data set 1, 2, 3 is 2. Suppose that you have the data set 1, 2, 297. The mean is now $1 + 2 + 297$ divided by 3, which is $300 \div 3 = 100$. However, the median of the data set 1, 2, 297 is still 2.

REMEMBER

Outliers affect the mean, but they don't affect the median. The mean gets pulled in the direction of the outlier and may not truly represent a "typical" value in the data set.

5. This problem gives you one way to check your answer to see if it makes sense.

   a. Because it averages out all the data in the set, the mean has to be somewhere between the largest and smallest values in the data set.

   b. The mean could equal the maximum value in a data set if all the values in the data set are the same; otherwise, any other value that isn't at the maximum pulls the mean down.

6. Many answers are possible. The key is to put the same number in the middle. One possible answer: data set 1: 100, 200, 300; data set 2: 199, 200, 201. The mean and median of both data sets is 200. These two data sets have the same center with totally different ranges (or spreads). If you want to tell the story about a data set, the center isn't enough because it can't distinguish between two data sets with different spreads.

7. This problem really points out what happens to the measures of center when you add any constant to all the values in the data set.

a. The mean also increases by $1,000 to $51,000, because you literally pick up all the salaries, move them up $1,000 on the number line, and put them back down, which moves the mean by the same amount.

b. The median also increases by the same amount, to $51,000, for the same reason.

TIP

Adding or subtracting a constant to or from all the values in a data set changes the mean and median by that same constant. Be careful — that constant could be negative as well as positive.

(8) This scenario highlights what happens when you multiply all the data by a constant. Here, the constant is 1.1, because you take the old salary, call it *X*, and add 10% of the *X* to it: $X + 0.10X$. But, $X + 0.10X = 1.10X$ so, in other words, 1.10 times the original salary gets you the new salary.

a. The mean also increases by 10% to become $\$50,000 * 1.1 = \$55,000$, because you multiply each value in the data set by 1.1.

b. The median also increases by 10% to become $55,000 for the same reason.

(9) The standard deviation can't be negative because of the squaring that goes on in its calculation. However, it can be 0, although it happens only when the data set has no deviation in it — in other words, when all the data are exactly the same value. For example, 1, 1, 1 or 2, 2, 2, 2, 2 are two data sets with a standard deviation of 0.

(10) If you choose 1, 1, 5, 5, you get the largest standard deviation possible, because these numbers are as far as possible from the mean (which is 3).

(11) Adding a constant to the data doesn't change the standard deviation, because you just relocate the data in a different spot on the number line; you don't change how far apart the values are from the mean.

(12) Multiplying by a constant changes the standard deviation. If you multiply an entire data set by 1.1, the spread increases. Suppose that two employees have salaries of $30,000 and $50,000 — right now, the figures are $20,000 apart. With a 10% raise, they become $33,000 and $55,000, making them $22,000 apart (the rich get richer, and the poor get less rich). If you recalculate the standard deviation, you find that it goes up here by a factor of 1.1 as well.

REMEMBER

The new standard deviation becomes *c* times the old standard deviation, when you multiply the data set by a nonnegative constant *c*. If you multiply the data by a negative constant, $-c$, the new standard deviation becomes $|c|$ times the old standard deviation (again, because of the squaring that goes on, the negative sign disappears). Also note that if *c* is a number between 0 and 1, the new standard deviation gets smaller than the old one.

(13) The median is 3.5 and divides the data into the lower half (1, 2, 3) and the upper half (4, 5, 6). The median of the lower half is Q1, which is 2, and the median of the upper half is Q3, which is 5. IQR is then $Q3 - Q1 = 5 - 2 = 3$.

(14) In this case, the median is 2, Q1 equals 2, and Q3 also equals 2, so the IQR is $2 - 2 = 0$. This makes sense because there is no spread in this data set anywhere, least of all in the middle 50 percent of the data.

(15) IQR is not affected by skewness or outliers because it measures only the range in the middle 50 percent of the data. It does not pay attention at all to the numbers on the outside edges, which is where skewness or outliers show up.

» Making sense of the information presented in charts and graphs

» Spotting misleading charts and graphs

Chapter **3**

Organizing Categorical Data: Charts and Graphs

Organizing categorical data involves listing each of the possible categories that the responses can take on, along with the number or percentage of individuals in each group. In this chapter, you practice making, interpreting, evaluating, and critiquing charts and graphs for categorical data.

Making, Interpreting, and Evaluating Pie Charts

A pie chart is a circle, or pie, whose slices show the percentage of individuals that fall into each category. Larger slices signify categories that include more individuals than the smaller slices. Before you make a pie chart, you can first summarize the data in table format. A frequency table shows how many individuals fall into each category (the sum of which is the *total sample size*). A relative frequency table shows what percentage of individuals fall into each category by taking the frequencies and dividing by the total sample size. The relative frequencies should sum to 1 or 100 percent (subject to possible round-off error).

Not all pie charts are correct, clear, and fair, however. Here are some common problems with pie charts that you should look out for:

- Important information is missing from the pie chart, such as the total number of individuals.
- The categories aren't broken down far enough; for example, a pie chart contains only three slices, and the biggest one is vaguely described as "other."
- The number of categories is so large that you can't really see anything (such as a pie chart where every individual represents a slice of the pie).

Gaining experience with pie charts helps you see some of the nuances that affect the credibility of their results. See the following for an example of interpreting and critiquing a pie chart.

EXAMPLE

Q. A hardware store wants to know what percentage of its customers are women. The manager takes a random sample of 76 customers who enter the store and records their gender. Twenty-two customers are females; the rest are males. I summarize the results in the following pie chart.

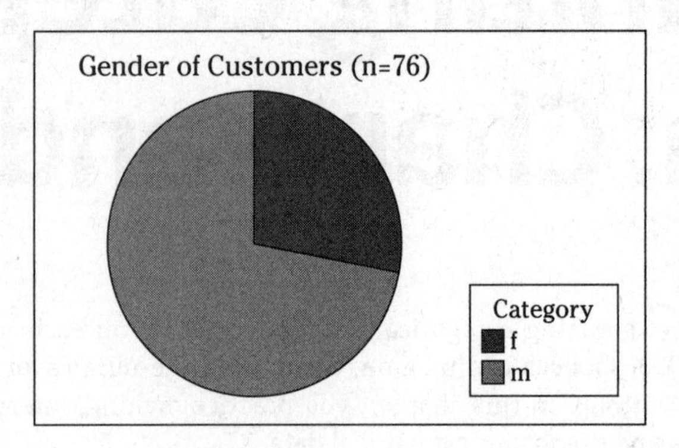

a. Describe the results.

b. How can this pie chart be improved?

A. Apparently, the DIY craze is popular with women, too.

a. The results of the pie chart show that the percentage of female customers appears to be around ⅓ (or around 33%).

b. You can improve the chart by showing the exact percentages in each slice. (The actual percentages are females: 28.9%; males: 71.1%.)

1 Suppose that 375 individuals are asked what type of vehicle they own: SUV, truck, or car. See the following frequency table.

a. Make a relative frequency table of these results.

b. Make a pie chart of these results.

c. Interpret the results.

Category	Frequency
SUV	150
Truck	125
Car	100
Total	375

2 Suppose that a restaurant owner keeps track of data on when his customers patronize his restaurant: breakfast, lunch, dinner, or other times. For a month, he takes time to check off which category each customer falls into. He records data on 1,000 customers for the month. The pie chart in the following figure shows his results.

a. What does this information tell the restaurant owner?

b. Can you spot a problem with the "other" category? How can this study be improved in the future?

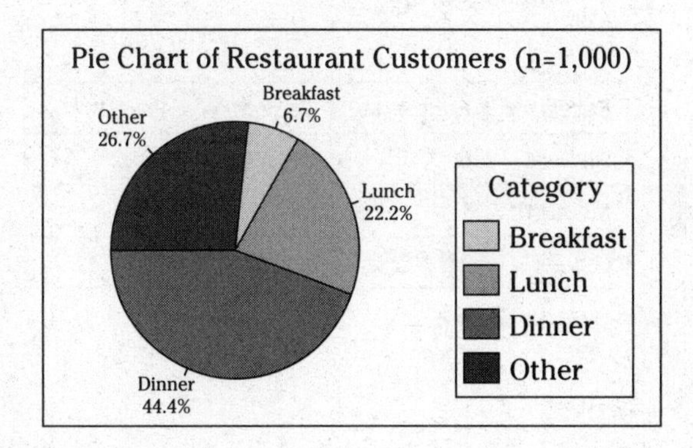

3 Suppose that an office manager wants to try to figure out a better way to multitask. She notices that answering emails is one of her most time-consuming duties, so she decides to categorize her emails into five groups: (1) highest priority; (2) medium priority; (3) very low priority; (4) personal; and (5) SPAM that she can delete immediately. You can see a frequency table of her results over a two-week period in the following table.

a. Make a relative frequency table of this data.

b. Make a pie chart of this data.

c. Interpret the results for the office manager.

Category	Frequency	Category	Frequency
Highest	60	Personal	50
Medium	120	SPAM	150
Lowest	20	Total	400

4 Suppose that a survey is conducted to see what types of pets people own. The survey of 100 adults finds that 40 of the people own a dog, 60 own a cat, 20 own fish, and 10 own some sort of rodent (hamster, gerbil, mouse, and so on). Can this data be organized in a pie chart? Explain your answer.

5 Suppose that as part of a driver's education program, students have to observe drivers in the real world and see how consistently they come to a complete stop at intersections. The students sit at an intersection for four hours and record whether each driver comes to a complete stop, rolls through the stop sign slowly, or runs the stop sign altogether. You can see the data from the study in the following pie chart.

a. Interpret the results as they appear on the pie chart.

b. What information is missing from the pie chart?

c. Does the missing information affect the interpretation of the results?

d. Should you make a generalization of all drivers based on this data?

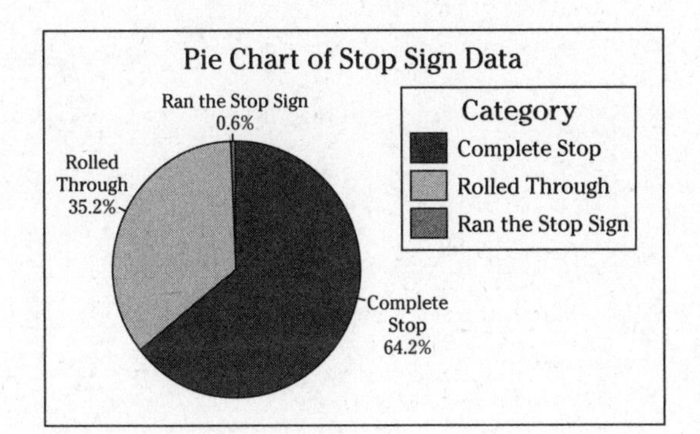

6 A survey is conducted to determine whether 20 office employees of a certain company would prefer to work at home, if given the chance. Of the 10 women surveyed, 7 say they would prefer to work at home, and 3 say no. Of the 10 men surveyed, 8 say no, and 2 say yes. Compare the results by using two pie charts. Does gender seem to affect one's preference to work at home? Explain.

7 Give an example of categorical data that you can't summarize correctly by using a pie chart.

8 Vault.com conducts a survey that asks employees and employers if they think surfing non-work-related websites compromises employee productivity. I summarize the results with the following pie charts.

a. Interpret these results.

b. What important information is missing from these pie charts?

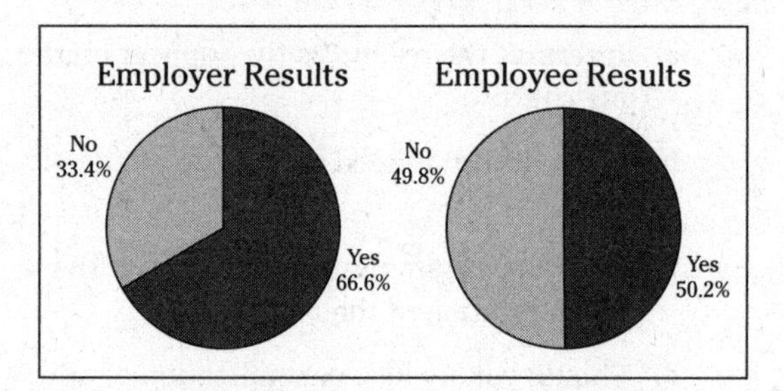

Making, Interpreting, and Evaluating Bar Graphs

A bar graph (or *bar chart*) is perhaps the most common data display used by the media. Like a pie chart, a *bar graph* breaks categorical data down by groups, showing the frequency or relative frequency for each group. A bar graph, however, uses bars of differing lengths to represent the number or percentage in each group.

Before making a bar graph, you may want to first organize the data into a frequency table or relative frequency table. Let each bar represent a category, and let the height of each bar represent the frequency or relative frequency for that category.

As with pie charts, not all bar graphs are fair, correct, and clear. You have to be very watchful of subtle changes in the graph that can greatly affect its appearance.

See the following for an example of comparing a bar graph to a pie chart.

EXAMPLE

Q. Following are a pie chart and bar graph (respectively) of scores from a quiz of ten questions, where the data shows the number of questions answered correctly. Name one advantage each has over the other.

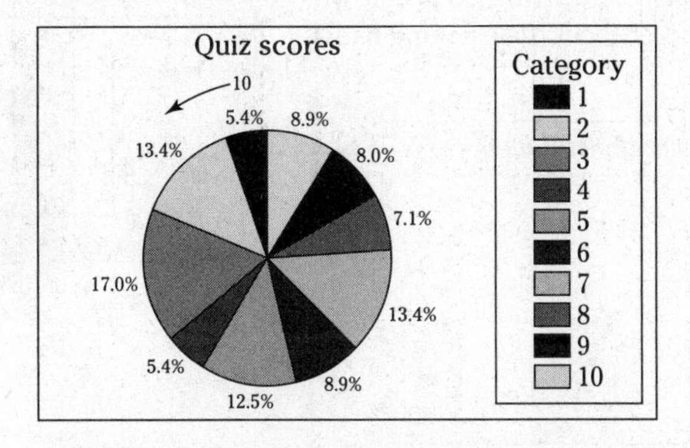

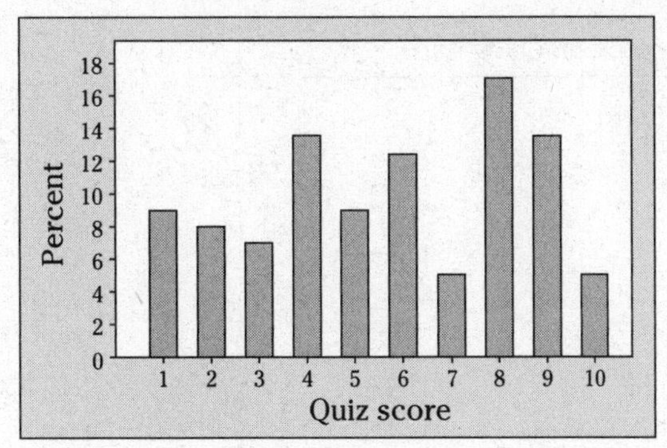

A. The pie chart shows everything as part of a whole, so you can make relative comparisons, and you know it all sums to 100%. The bar graph, however, makes it easier to compare the groups to each other. (And if the pie chart doesn't show the percentages, you have a much harder time estimating the percent in each group.)

9 The following figure shows a frequency bar graph of 500 people who make up three categories (1: in favor of a smoking ban; 2: against a smoking ban; and 3: no opinion).

a. Make a relative frequency table of this data.

b. Use the relative frequency table to make a bar graph of this data.

c. Interpret the results. (How do people in the sample feel about the smoking ban?)

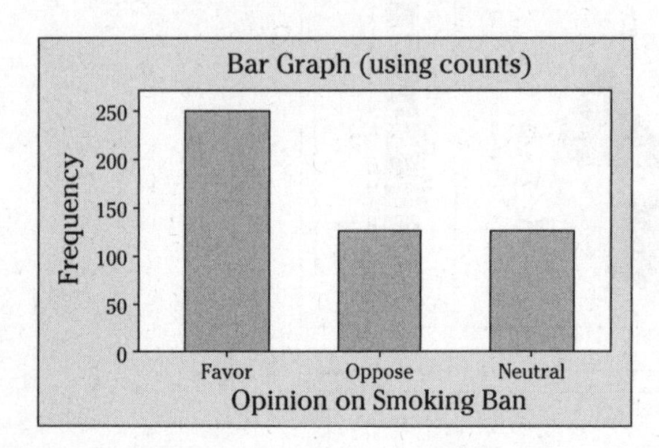

10 Suppose that a health club asks 30 customers to rate the services as very good (1), good (2), fair (3), or poor (4). You can see the results in the following bar graph. What percentage of the customers rated the services as good?

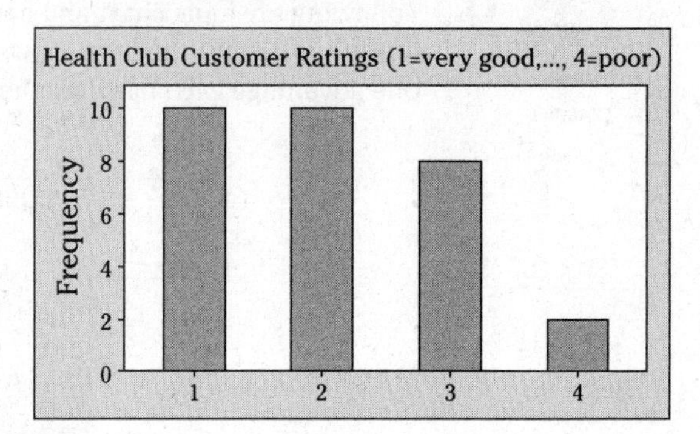

11 A polling organization wants to find out what voters think of Issue X. It chooses a random sample of voters and asks them for their opinions of Issue X: yes, no, or no opinion. I organize the results in the following bar graph.

a. Make a frequency table of these results (including the total number).

b. Evaluate the bar graph as to whether it fairly represents the results.

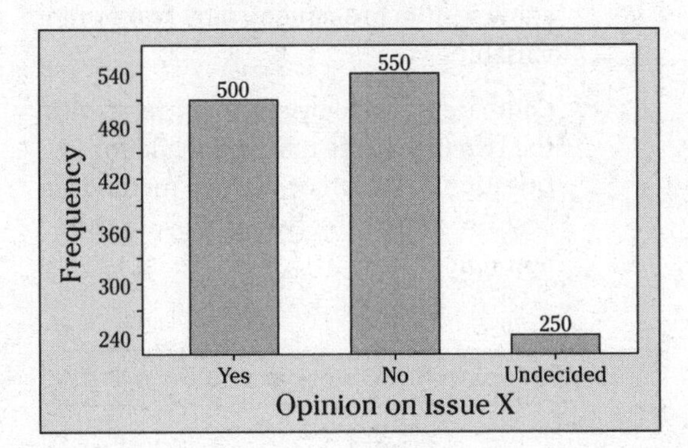

12 Suppose that a random sample of 270 graduating seniors are asked what their immediate priorities are, including whether buying a house is a priority. The results are shown in the following bar graph.

a. The bar graph is misleading; explain why.

b. Make a new bar graph that more fairly presents the results.

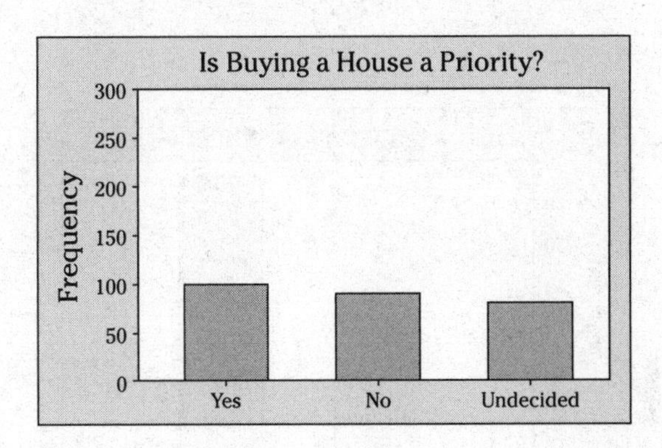

13 A car dealer specializing in minivan sales conducts a survey to find out more about who its customers are. One of the variables the company measures is gender; the results of this part of the survey are shown in the following bar graph.

a. Interpret these results.

b. Explain whether you think the bar graph is a fair and accurate representation of this data.

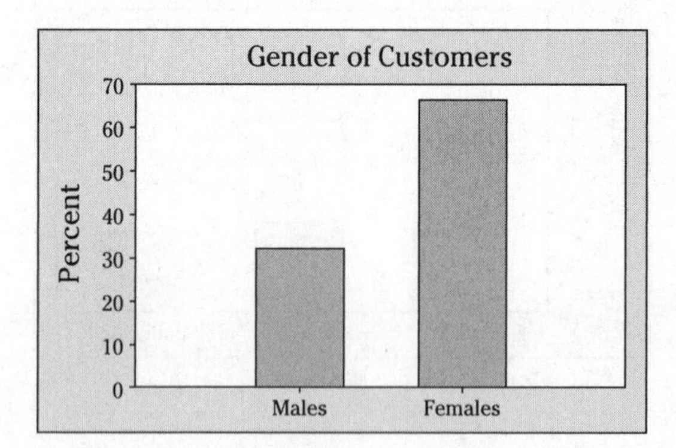

14 A survey is conducted to determine whether 20 office employees of a certain company would prefer to work at home, if given the chance. The overall results are shown in the first bar graph, and the results broken down by gender are presented in the second.

a. Interpret the results of each graph.

b. Discuss the added value in including gender in the second bar graph. (The second bar graph in this problem is called a *side by side* bar graph and is often used to show results broken down by two or more variables.)

c. Compare the side by side bar graph with the two pie charts that you made for Question 6. Which of the two methods is best for comparing two groups, in your opinion?

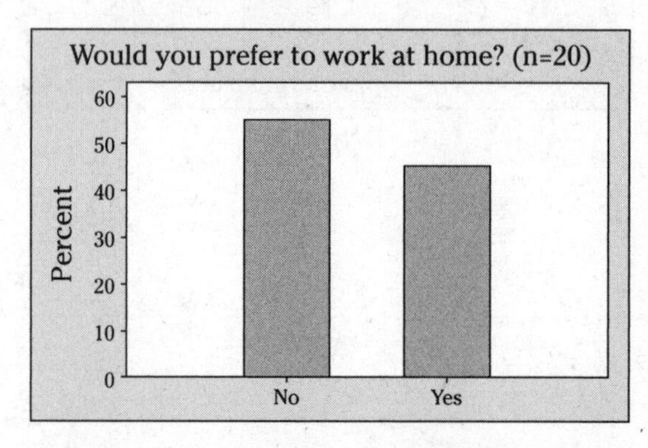

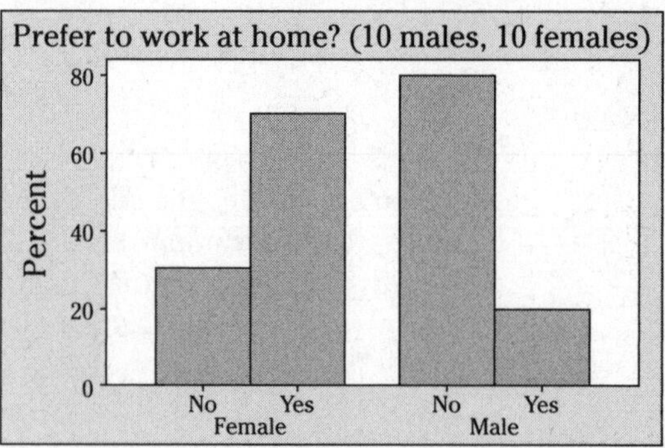

Answers to Problems in Organizing Categorical Data

1 Organizing the data in a relative frequency table before you make a pie chart is often helpful.

a. See the following relative frequency table.

Category	Relative Frequency
SUV	$150 \div 375 = 0.400$ or 40.0%
Truck	$125 \div 375 = 0.333$ or 33.3%
Car	$100 \div 375 = 0.267$ or 26.7%
Total	$375 \div 375 = 1.00$ or 100%

b. The following pie chart shows the results using Minitab. If you make your pie chart by hand, it should look similar, but it may not be exactly the same because it can be hard to gauge how big the slices should be.

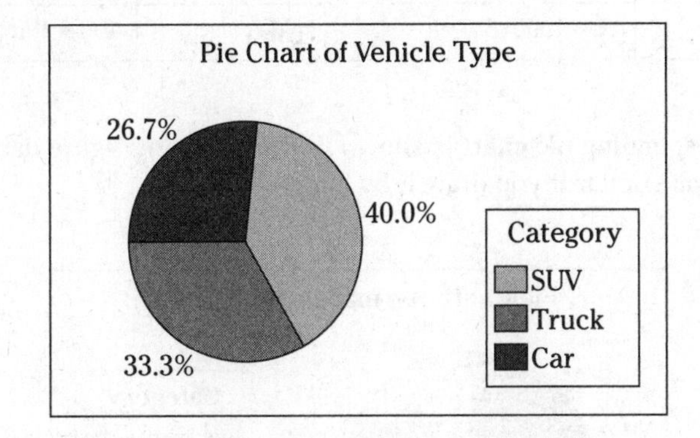

Although making pie charts by hand works fine (and on exams, you'll likely have to do so), using a computer is the easiest and most accurate way to make a pie chart (using programs such as Excel or Minitab). The problem is getting the size of the slices just right. You can take the percentage for the category and multiply by 360 degrees to figure out how big of an angle to make, if you remember how to do that sort of thing from trigonometry. Or you can divide the pie into quarters (25% each) and divide each quarter into eighths (12.5% each), using dotted lines, to make your best estimate from there. Remembering that half the pie chart represents 50% can be helpful, too.

c. From these results, you can see that 40% of the individuals have SUVs. Trucks and cars split up the remainder of the group, with 26.7% owning cars and the rest owning trucks.

2. Watch out for pie charts that have large, ambiguous "other" categories.

   a. The pie chart shows that, of the three meals, dinner brings in the most business (44% compared to only 22% for lunch and around 7% for breakfast).

   b. The pie chart does have a problem. The "other" group is very large, comprising almost 27% of the business, but he has no idea when those people come in, so he has little information to work from. He may have received better results if he had broken the "other" category into more categories, such as between breakfast and lunch, between lunch and dinner, after dinner, bakery purchases on the go, and so on.

WARNING

Beware of slices of the pie labeled "other" or "miscellaneous" that become larger than many of the other slices. This discrepancy is a clue that the creator should've added more categories.

3. Pie charts do a nice job of summarizing data accurately and quickly.

   a. It didn't take long for the office manager to realize that SPAM floods her inbox and crowds out the more important emails. The relative frequency table explains why.

Category	Relative Frequency	Category	Relative Frequency
Highest	$60 \div 400 = 15\%$	Personal	$50 \div 400 = 12.5\%$
Medium	$120 \div 400 = 30\%$	SPAM	$150 \div 400 = 37.5\%$
Low	$20 \div 400 = 5\%$	**Total**	$400 \div 400 = 100\%$

   b. The corresponding pie chart is shown in the following figure using Minitab. Your pie chart should look similar if you draw it by hand.

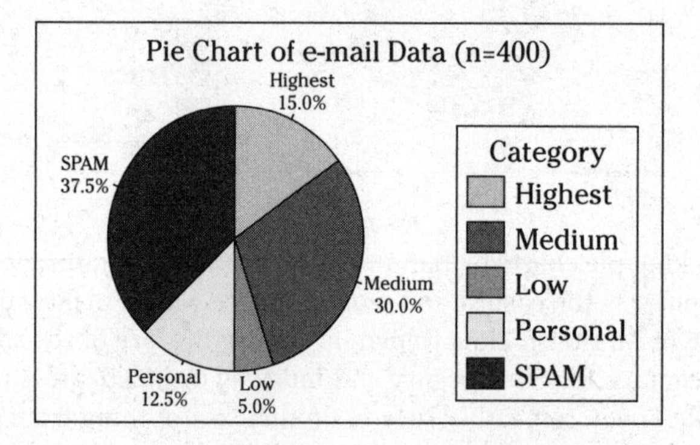

TIP

If you plan to draw a pie chart by hand, try starting with the largest slice and working your way down. Your results should look similar to charts drawn by any computer software package. If you use Microsoft Excel, you need to make a table first and the pie chart second.

   c. The results tell the office manager that she gets a great deal of SPAM. She also gets quite a bit of personal email, which she can save for breaks and lunch to maximize her work time. She also sees that $15\% + 30\% = 45\%$ of her emails are high to medium priority, which can cause some stress.

4. No. The data appears to report $40+60+20+10=130$ pet owners, but the total number of people surveyed was only 100. Why? Because some folks are counted more than once if they own more than one different type of pet. The total doesn't add up to *n* (the sample size), and the percents don't add up to 100 if you divide each frequency by *n* (which is what you should do). Therefore, a pie chart doesn't work for this survey. (A bar chart is a good alternative.)

All the percentages in a pie chart must add up to 100% or close to it (subject to round-off error).

5. Interpreting pie charts can seem so easy that you may be tempted to go too far at times.

   a. The pie chart shows that 64.2% of the drivers who approached the intersection came to a complete stop, 35.2% rolled through the stop sign, and 0.6% (or 0.006) actually ran the stop sign.

   b. You have no indication of how many cars the students examined (*n*, the sample size, isn't known). They may have seen a small number of vehicles.

   c. Not knowing the sample size upon which a pie chart is based can lead to imprecise or even misleading results.

   d. The data came from only one intersection on a single day for a four-hour period, so you can't make generalizations to all drivers from this very limited data set.

Look for the total sample size, which is related to the precision of your results.

6. The pie charts are shown in the following figure. Yes, gender does seem to be related to the preference to work at home (for this company). More females at this company prefer to work at home (70%) than males (20%).

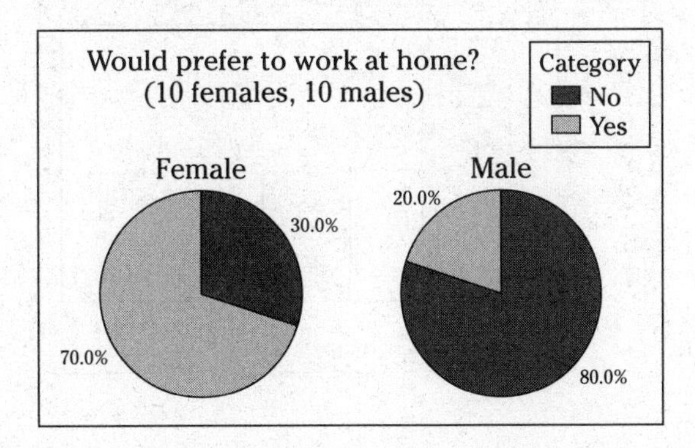

7. Any example where the percentages don't sum to 100 — where an individual or object can be in more than one group at the same time. For example, suppose that a group of adults are asked what kinds of activities they like to do on a Friday night: 41% say watch TV, 50% say go to a movie, and 60% say go out to eat. The surveyor doesn't ask what the people like to do best, so they can choose more than one at a time. A pie chart doesn't make sense here.

8. A missing sample size is a common error with graphs and charts. (***Note:*** Vault.com did provide the missing information; I just removed it to see if you'd notice.)

   a. More employers feel employee surfing compromises productivity by a 2-to-1 margin (67% yes to 33% no). But employees are equally split on the issue (about 50% yes to 50% no). That's not surprising, is it?

   b. The sample sizes are missing. You don't know whether 10 people or 10,000 people responded, which affects the precision of the results. The date of the survey is also missing. (Turns out the Internet company surveyed 451 employees and 670 employers in the fall of 2000.)

REMEMBER

A pie chart should stand alone; all the necessary information should be included and labeled within the chart.

9. Relative frequencies (percents) allow you to make easy comparisons between groups.

   a. See the following relative frequency table.

Category	Relative Frequency
Support smoking ban	$250 \div 500 = 50\%$
Oppose smoking ban	$125 \div 500 = 25\%$
No opinion	$125 \div 500 = 25\%$
Total	$500 \div 500 = 100\%$

   b. The following figure shows the bar graph.

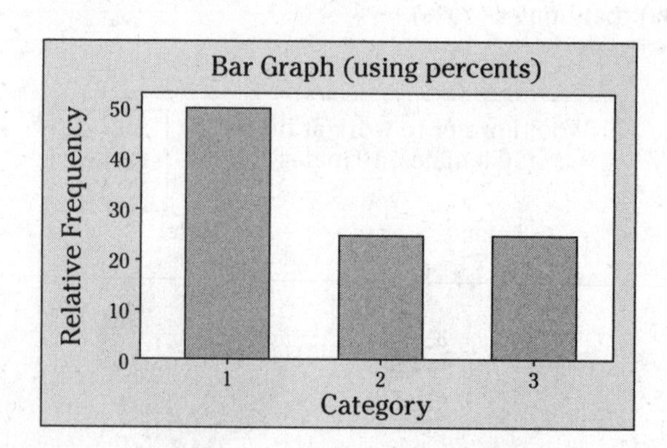

   c. Half of the individuals support the smoking ban, and the rest of the people are evenly split between opposition and no opinion.

10. The bar graph shows that 10 out of 30 customers ($10 \div 30 = 33.3\%$) rated the services as good. Notice that the answer isn't 10%, because frequencies appeared on the *y*-axis, not relative frequencies.

REMEMBER

When interpreting the results of a bar graph, be sure that you know what the graph is meant to report: counts (frequencies) or percents (relative frequencies).

11 Starting point, scale, and sample size are three major ways a bar graph can mislead you.

a. See the following frequency table of the results of the opinion poll on Issue X. The frequencies come from the bar graph and represent the height of each bar. A total of 1,300 people were surveyed.

Opinion	Frequency
Yes	500
No	550
No opinion	250
Total	1,300

b. The bar graph appears to be misleading. Notice that the bar for the "undecided" group is much less than half of the length of the bar for the "yes" group, because the frequency axis starts at 240 and not at 0. The following figure shows a better bar graph. The scale on this second bar graph is a little different (it uses increments of 100 rather than 30, which appear on the original bar graph). The new scale makes the graph easier to read, because it results in "nice" numbers on the tick marks like 300 and 400. The scale changes quite a bit, but the biggest problem with the original bar graph isn't the scale; it's the starting point. The total sample size is now included on the x-axis, making the results easier to interpret.

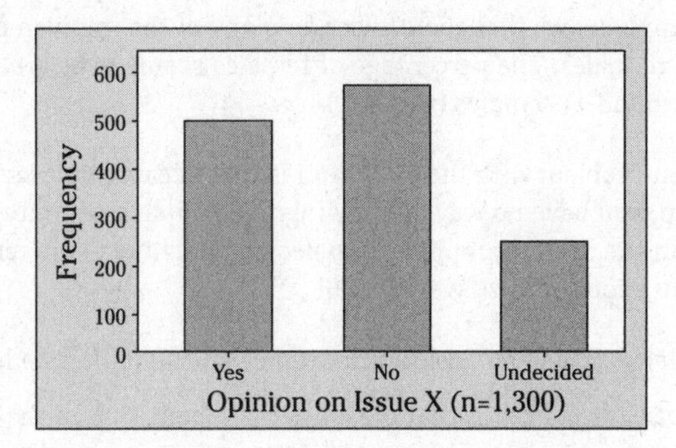

TIP

Watch the starting point on the counts/percents axis. If it doesn't start at 0, the differences in the bar lengths may appear larger than they really are.

12 Scale and ending-point errors can squeeze the space in between bars and create extra, unnecessary space in the graph. Each makes the graph misleading.

a. The bar graph is misleading because it makes the differences in the bars look smaller than they truly are. You can attribute this deception to two things. First, the scale on the y-axis is 50, which is bigger than it should be, squeezing the bars closer together in length and making differences appear relatively small. Also, the y-axis goes all the way to 300, when it could easily stop at half that distance. This creates a large portion of unused space in the graph, again making the lengths of the bars look smaller than they should be. The extra space also makes it appear that the sample size is smaller than it actually is (like not very many people were in each group).

b. A more fair and informative bar graph is shown in the following figure.

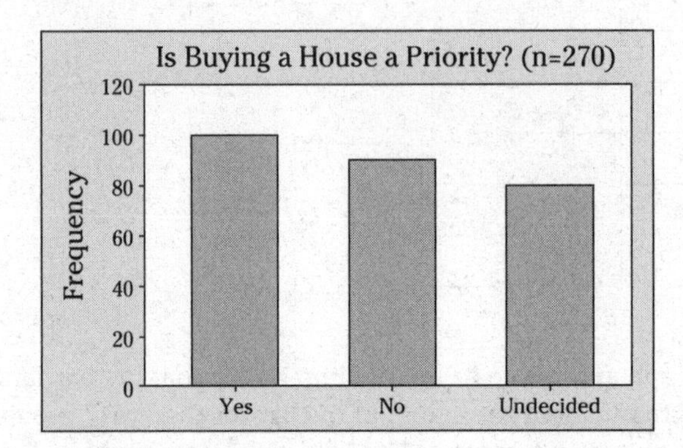

TIP

Watch the scale (size of the increments or tick marks) on the counts/percents axis. Look for scales that show differences as less or more dramatic than they should be.

13 Although percents allow you to make relative comparisons, you still need to have the sample sizes to determine how precise the results are.

a. The bar graph shows that about twice as many of the minivan customers were females compared to males. The percentage of females seems to be around 67% (about ⅔), compared to around 33% males (about ⅓).

b. The biggest problem with the bar graph is that because you see only the percentages in each group, you have no way of knowing the sample size. Sixty-seven percent females could mean the dealer sampled 3,000 people and 2,000 were female, or it could mean he sampled 30 people and 20 were female.

14 A second variable may be of critical importance and shouldn't be left out.

a. The first bar graph in Question 14 shows that about 55% of the office employees of this company prefer to work at home, and 45% enjoy an at-work environment. The second graph shows that, of the females, 70% prefer to work at home, and 30% wouldn't. (Notice the percents sum to 100 for each group separately, allowing you to make comparisons between groups.) Of the males, only 20% would prefer working at home, and 80% wouldn't.

b. The second bar graph is more interesting, because it shows that results differ depending on gender.

c. The side-by-side bars are my personal choice over the two pie charts, because the former are shown using the same scale, making it easier to visually see the differences. Pie charts usually have different slices for all the groups, making it hard to compare them without an obvious difference.

» Recognizing misleading histograms

» Making and interpreting box plots

» Understanding line graphs

Chapter 4
Organizing Quantitative Data: Charts and Graphs

Quantitative data comes to you in the form of measurements of some kind, where the numbers make sense as numbers (and not as placeholders for categories). Statisticians use different types of charts and graphs to organize quantitative data; in this chapter, you experience histograms, box plots, and line graphs (also known as time charts.) Depending on your situation, you can use them to look at how the data are distributed among the values, where the "middles" of the data are, how spread out the data are, or how they look over time — all important characteristics of a data set.

Even though you may not see all the graphs and charts in this chapter put out by the media, you see them with great frequency in your statistics class, so you need to be very comfortable with all of them and their interpretation. This chapter helps you settle in.

Creating a Histogram

A *histogram* is a bar graph made for quantitative data. Because the data are numerical, you divide it into groups without leaving any gaps in between (so the bars are connected). The Y-axis shows either frequencies (counts) or relative frequencies (percents) of the data that fall into each group (see Chapter 1 for more on these topics).

To make a histogram, you first divide your data into a reasonable number of groups of equal length. Tally up the number of values in the data set that fall into each group (in other words, make a frequency table). If a data point falls on the boundary, make a decision as to which group to put it into, making sure you stay consistent (always put it in the higher of the two, or always put it in the lower of the two). Make a bar graph, using the groups and their frequencies — a *frequency histogram.* If you divide the frequencies by the total sample size, you get the percentage that falls into each group. A table that shows the groups and their percents is a relative frequency table. The corresponding histogram is a *relative frequency histogram.*

TIP

I created all the histograms in this workbook by using Minitab. You can use a different software package, or you can make your histograms by hand. Either way, your choice of interval widths (called bins by computer packages) may be different from the ones I choose, which is fine, as long as yours look similar. And they will, as long as you don't use an unusually low or high number of bars and your bars are of equal width. You may also choose different start/end points for each interval, and that's fine as well. Just be sure to label everything clearly so your instructor can see what you're trying to do. And be consistent about values that end up right on a border; always put them in the lower grouping, or always put them in the upper grouping. Or do like I do: Make your bin cutoffs non-overlapping to avoid the problem. If you do have a choice, however, make your histograms by using a computer package like Minitab. It makes your task much easier.

See the following for an example of making the two types of histograms.

EXAMPLE

Q. Test scores for a class of 30 students are shown in the following table.

Scores	Frequency
70–79	8
80–89	16
90–99	6

a. Make a frequency histogram.

b. Find the relative frequencies for each group.

c. Without actually drawing it, how would the relative frequency histogram compare to the frequency histogram?

A. Frequency histograms and relative frequency histograms look the same; they're just done using different scales on the Y-axis.

a. The frequency histogram for the scores data is shown in the following figure.

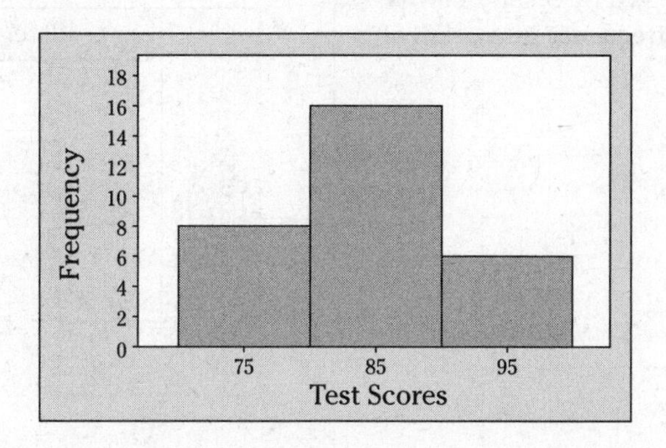

b. You find the relative frequencies by taking each frequency and dividing by 30 (the total sample size). The relative frequencies for these three groups are $8 \div 30 = 0.27$ or 27%; $16 \div 30 = 0.53$ or 53%; and $6 \div 30 = 0.20$ or 20%, respectively.

c. A histogram based on relative frequencies looks the same as the histogram (of the same data). The only difference is the label on the Y-axis.

---

1 You lose information from the data when you create a histogram. What information is lost?

2 Make a histogram from this data set of test scores: 72, 79, 81, 80, 63, 62, 89, 99, 50, 78, 87, 97, 55, 69, 97, 87, 88, 99, 76, 78, 65, 77, 88, 90, and 81. Would a pie chart be appropriate for this data?

3 Suppose that you take a survey of 45 homeowners to find out how many televisions they own. After you finish, you find that 2 people own no TVs, 17 people own one, 22 people own two, 3 own three, and 1 owns four. Make a relative frequency histogram of this data.

4 Suppose that you have a loaded die. You roll it several times and record the outcomes, which are shown in the following figure.

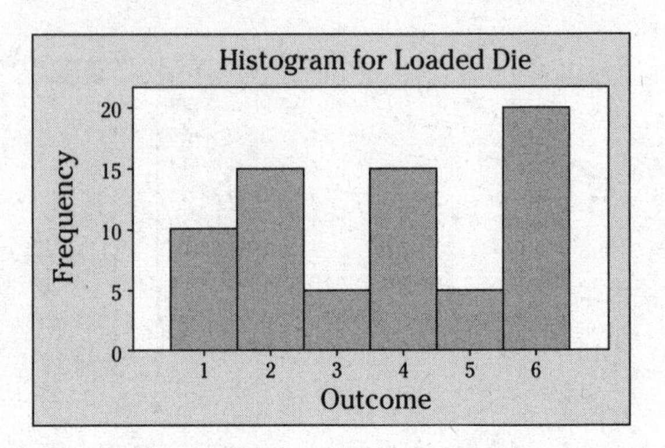

a. Make a relative frequency histogram of these results.

b. You can make a relative frequency histogram from a frequency histogram; can you go the other direction?

Making Sense of Histograms

A histogram gives you general information about three main features of your quantitative (numerical) data: the shape, center, and spread.

The *shape* of a histogram is shown by its general pattern. Many patterns are possible, and some are common (see Figure 4-1), including the following:

- **Bell-shaped:** Looks like a bell — a big lump in the middle and tails that go down on each side at about the same rate. (See Figure 4-1a.)
- **Right skewed:** A big part of the data is set off to the left, with a few larger observations trailing off to the right. (See Figure 4-1b.)
- **Left skewed:** A big part of the data is set off to the right, with a few smaller observations trailing off to the left. (See Figure 4-1c.)
- **Uniform:** All the bars have a similar height. (See Figure 4-1d.)
- **Bimodal:** Two peaks, or *modes.* (See Figure 4-1e.)
- **U-shaped:** Bimodal with the two peaks at the low and high ends, with less data in the middle. (See Figure 4-1f.)
- **Symmetric:** Looks the same on each side when you split it down the middle; bell-shaped, uniform, and U-shaped histograms are all examples of symmetric data. (See Figures 4-1a, d, and f.)

You can view the *center* of a histogram in two ways. One is the point on the *x*-axis where the graph balances, taking the actual values of the data into account. This point is called the *average*, and you can find it by locating the balancing point (imagine the data are on a teeter-totter). The other way to view center is locating the line in the histogram where 50 percent of the data lies on either side. The line is called the *median*, and it represents the physical middle of the data set. Imagine cutting the histogram in half so that half of the area lies on either side of the line.

Spread refers to the distance between the data, either relative to each other or relative to some central point. One crude way to measure spread is to find the *range*, or the distance between the largest value and the smallest value (see Chapter 2). Another way is to look for the average distance from the middle, otherwise known as the *standard deviation.* The standard deviation is hard to come up with by just looking at a histogram, but you can get a rough idea if you take the range divided by 6. If the heights of the bars close to the middle seem very tall, that means most of the values are close to the mean, indicating a small standard deviation. If the bars appear short, you may have a larger standard deviation.

You can do actual summary statistics to calculate the quantitative data (see Chapter 2), but a histogram can give you a general direction for finding these milestones. And like pie charts and bar graphs (see Chapter 3), not all histograms are fair, complete, and accurate. You have to know what to look for to evaluate them.

See the following for an example of catching up to your data by using a histogram.

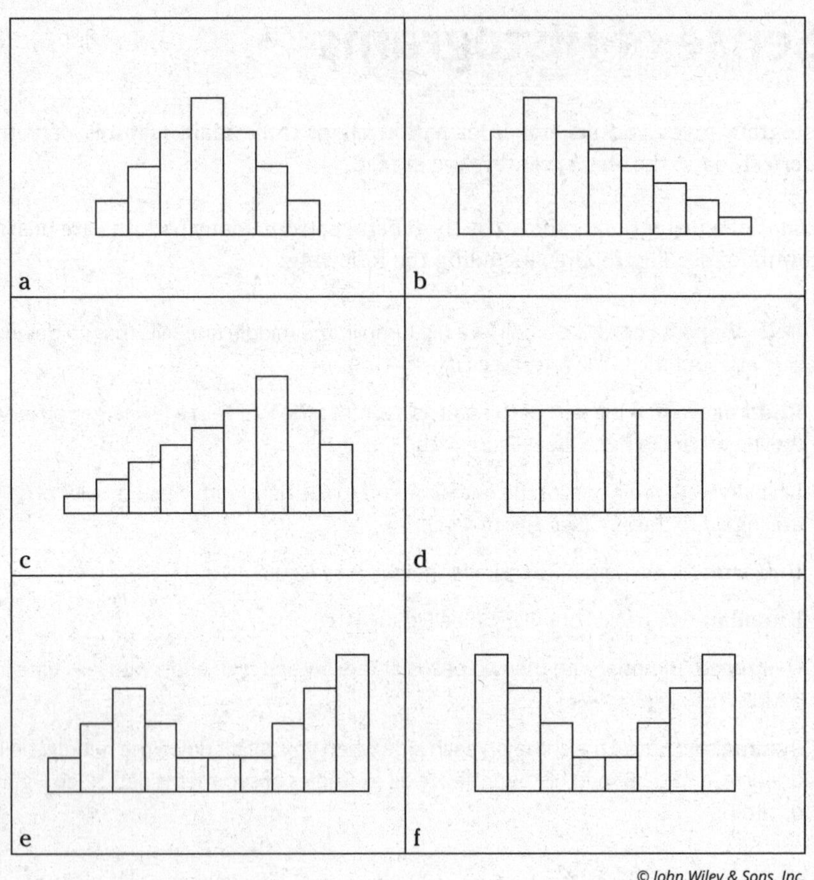

FIGURE 4-1: Histograms have several common patterns.

© John Wiley & Sons, Inc.

EXAMPLE

Q. The police checked the speeds of cars after the city painted lines on a certain section of street where the road narrows. The speeds are organized in the histogram shown in the following figure. Describe what this histogram tells you about the speeds of the cars.

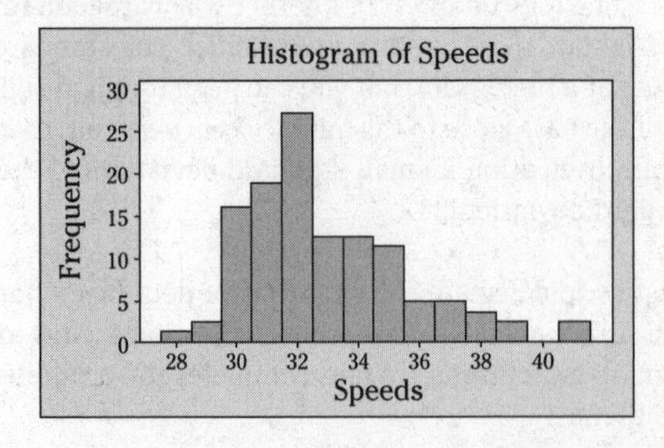

A. The speeds of the cars in this data set range from 28 mph (lowest) to 41 mph (highest). Most of the cars traveled from 30 to 35 mph (you can tell by noting that those bars have the highest frequencies). A few cars drove faster than the rest (noted by the few short bars at the upper end), which indicates a skewed right shape. The average speed seems to center around 32 mph. (But this is hard to tell without doing more analysis.)

5 An ATM machine asks customers who use the "fast cash" option to choose an amount in $50 increments from $100 to $500. Results from a recent sample of customer withdrawals are shown in the following figure. Discuss the shape, center, and spread of the data.

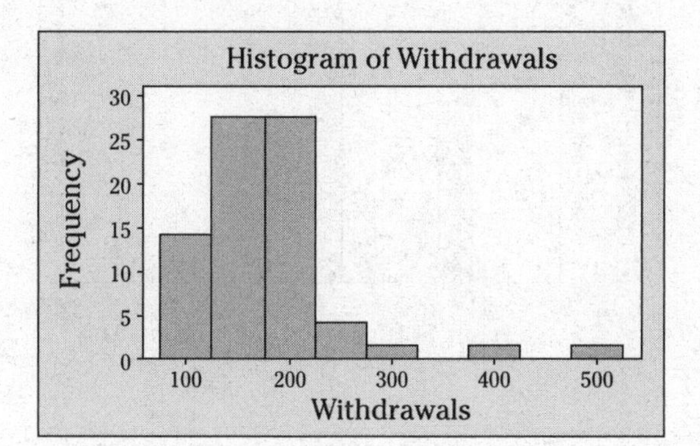

6 A histogram of a sample of rods made by Rowdy Rods is shown in the following figure. The rods should be 100 inches in length. Discuss the company's accuracy (in terms of meeting the length specification) by interpreting the shape, center, and spread of the data.

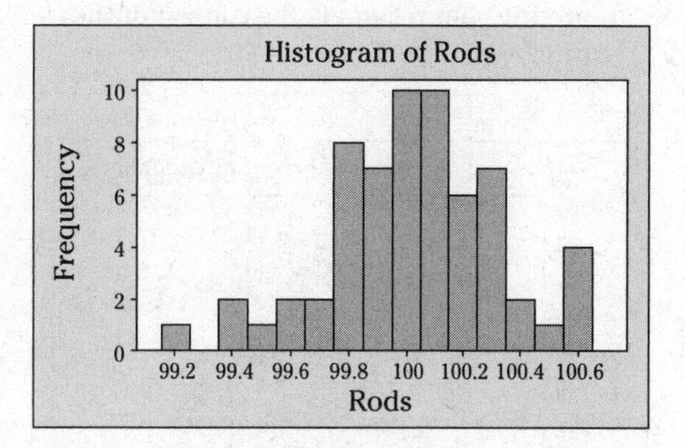

7 A histogram of the amount of money 317 households spent on fruits and vegetables in a year is shown in the following figure (based on a random sample). Discuss the shape, center, and spread. What do these three characteristics say about how much families spend on fruits and vegetables? (Such analysis is called interpreting your results in the context of the problem.)

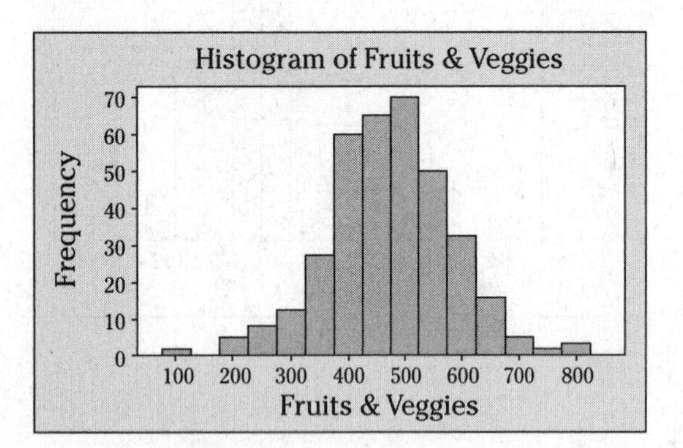

8 An investor monitors the percentage return for a particular group of stocks in Portfolio A over a year period. The one-year percentage returns for these stocks are shown in the following figure.

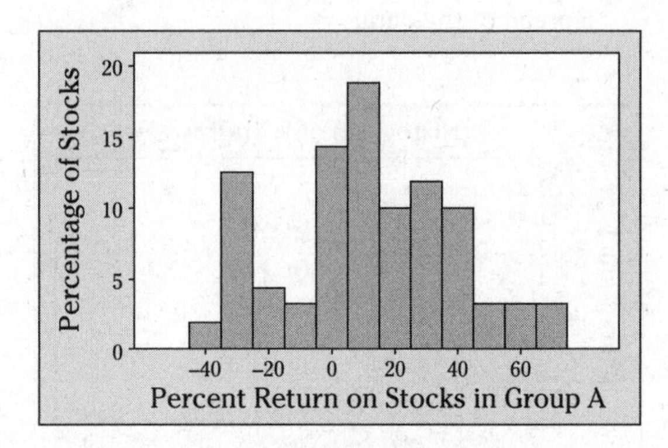

a. Some of the values on the x-axis are negative numbers. What does this mean?

b. On any histogram in general, when (if ever) can the x- and/or y-axis contain negative values?

9 The incomes of last year's new graduates of a certain large and very successful program are shown in the following figure.

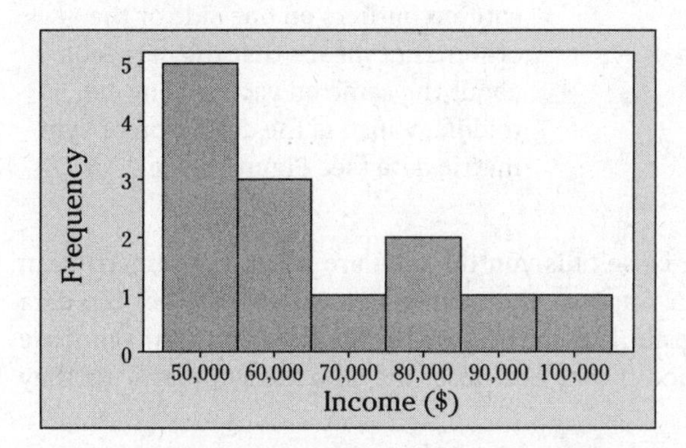

a. Discuss the implications for graduates of this program.

b. Estimate where the median salary is in this data set.

c. Do you see any issues that anyone who tries to interpret this data should take into account?

10 You make two histograms from two different data sets (see the following figures), each one containing 200 observations. Which of the histograms has a smaller spread: the first or the second?

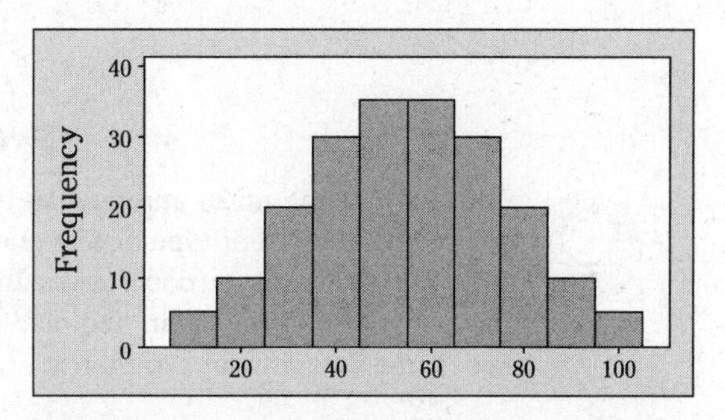

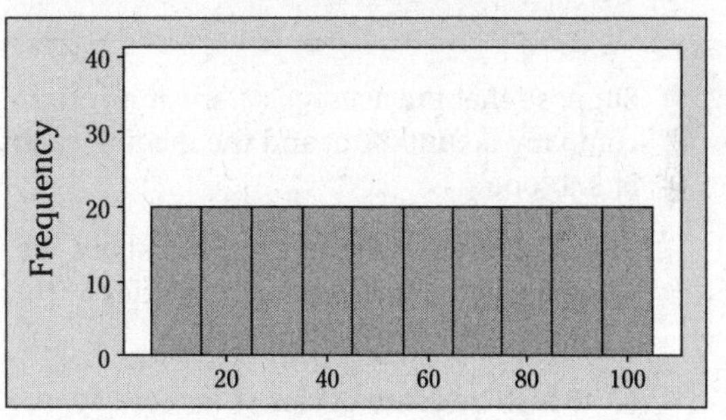

Straightening Out Skewed Data

You need to make special considerations for skewed data sets, in terms of which statistics are the most appropriate to use and when. You should also be aware of how using the wrong statistics can provide misleading answers.

The following example shows how you can relate the mean and median to learn about the shape of your data.

EXAMPLE

Q. Explain why a data set having the mean and the median close to being equal will have a shape that is roughly symmetric.

A. The mean is affected by outliers in the data, but the median is not. If the mean and median are close to each other, the data aren't skewed and likely don't contain outliers on one side or the other. That means that the data look about the same on each side of the middle, which is the definition of symmetric data (see Figure 4-1 a, d, or f).

TIP

The fact that the mean and median being close tells you the data are roughly symmetric can be used in a different type of test question. Suppose that someone asks you whether the data are symmetric, and you don't have a histogram, but you do have the mean and median. Compare the two values of the mean and median, and if they are close, the data are symmetric. If they aren't, the data are not symmetric.

---

11 Suppose that the average salary at a certain company is $100,000, and the median salary is $40,000.

a. What do these figures tell you about the shape of the histogram of salaries at this company?

b. Which measure of center is more appropriate here?

c. Suppose that the company goes through a salary negotiation. How can people on each side use these summary statistics to their advantage?

Suppose that you know that a data set is skewed left, and you know that the two measures of center are 19 and 38. Which figure is the mean and which is the median?

13 Can the mean of a data set be higher than most of the values in the set? If so, how? Can the median of a set be higher than most of the values? If so, how?

14 Is the standard deviation affected by skewed data? If so, how?

Spotting a Misleading Histogram

Readers can be misled by a histogram in ways that aren't possible with a bar graph. Remember that a histogram deals with numerical data, not categorical data (see Chapter 3), which means you have to determine how you want the numerical data broken down into groups to display on the horizontal axis. And how you determine those groupings can make the graph look very different. Watch for histograms that use scale to mislead readers. As with bar graphs, you can exaggerate differences by using a smaller scale on the vertical axis of a histogram, and you can downplay differences by using a larger scale.

See the following for an example of how people can make data misleading by the way they create their histograms.

Q. In Question 4, you see data from a die that's clearly loaded (not fair). However, someone (the gambler's lawyer, perhaps?) could make that same die appear fair by setting up the histogram a certain way. Explain how the following histogram (made from that same data) makes the die appear to be fair.

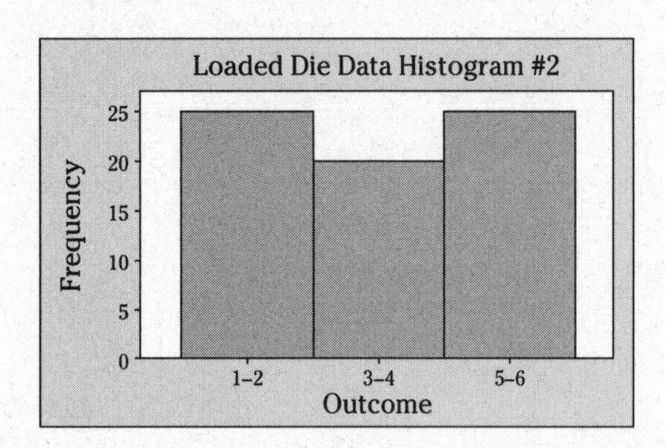

A. This histogram combines the results of rolling a 1 and 2, rolling a 3 and 4, and rolling a 5 and 6, so there are 3 bars on the histogram now, not 6. This histogram is misleading because when you combine the results into three groups of two outcomes each, the differences that make the die loaded don't show up. The lack of precision works to the advantage of the person who created the graph.

15 Suppose that your friend believes his gambling partner plays with a loaded die (not fair). He shows you a graph of the outcomes of the games played with this die (see the following figure). Based on this graph, do you agree with this person? Why or why not?

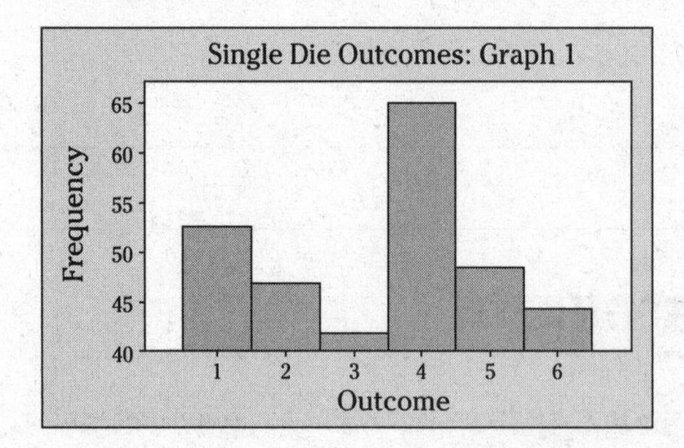

16 The first month's telephone bills for new customers of a certain phone company are shown in the following figure. The histogram showing the bills is misleading, however. Explain why, and suggest a solution.

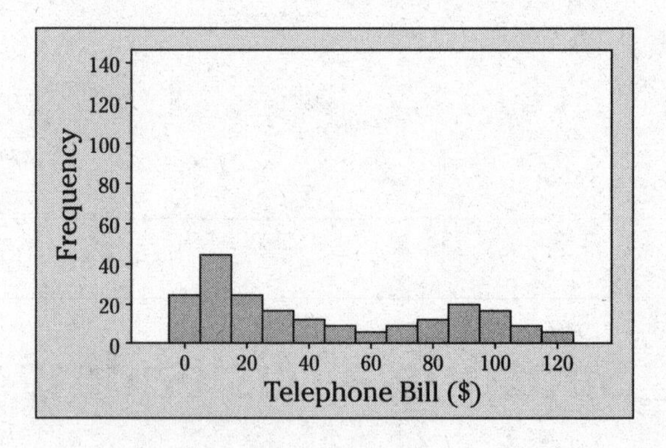

Making Box Plots

A *box plot* is a one-dimensional graph made for quantitative data that divides the data into four parts, each part containing 25 percent of the data. It is a graph of what is called the *five-number summary*, which consists of the minimum value, the first quartile (Q1; also known as the 25th percentile), the median, the 3rd quartile (Q3; also known as the 75th percentile), and the maximum value. Note that ordering the data set is important before finding these values. (See Chapter 2 for information on percentiles and all the other numbers in the five-number summary.)

To make a box plot, you find the five-number summary and graph its values just above a one-dimensional number line. Then draw a box around the area representing the first and third quartiles, draw a line through the box where the median is, and then draw lines out to the minimum and maximum values.

TIP

I created all the box plots in this workbook by using Minitab. You can use a different software package, or you can make your box plots by hand.

See the following for an example of making box plots.

EXAMPLE

Q. Make a box plot of the five-number summary indicated by 5, 7.5, 20, 27.5, 30 (where 5 is the minimum value, 7.5 is the first quartile, and so on).

A. This box plot is a vertical box plot with the 5 main values shown; the box surrounds the Q1 (7.5) to Q3 (27.5) area, the median is inside the box (20), and the lines go out to the min (5) and max (30). It's got a little more space in the bottom of the box and less in the top part, so we say it is skewed to the left (more area in the left/bottom side of the graph).

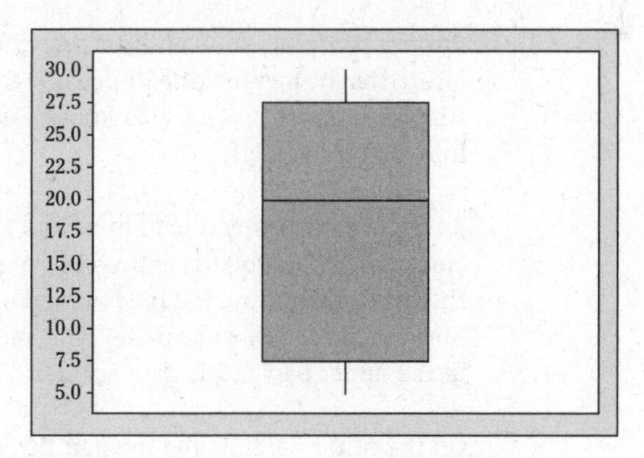

17 Make a box plot from the five-number summary: 3, 4, 7, 16, 17.

18 Make a box plot from the five-number summary: 100, 105, 120, 135, 140.

Interpreting Box Plots

From a box plot, you can interpret the shape, center, and spread of a data set. The center is located at the median, the line inside the box; the spread is the difference between the third and first quartiles (Q3 – Q1), also known as the interquartile range, or IQR (see Chapter 2 for more information on IQR).

As for shape, you can look at where the median lies within the box itself. If the median lies closer to the first quartile, then there is a smaller range in the 25 percent of the values between the first quartile and the median and a larger range in the 25 percent of the values between the median and the third quartile. That says that the data are skewed to the right (a larger distance in the upper part of the data set).

On the other hand, if the median lies closer to the third quartile, this means that the 25 percent of the data on the lower end of the box have a larger range (from the first quartile or the median), and 25 percent of the data in the upper end of the box have a smaller range (from the median to the third quartile). This says that the data are skewed left. If the median lies pretty much in the middle of the box, the data are fairly symmetric. Looking at the lines going out to the minimum and maximum values, you can spot outliers if they exist (values far from the rest).

The following is an example of interpreting a box plot.

EXAMPLE

Q. Interpret the shape, center, and spread of the following box plot.

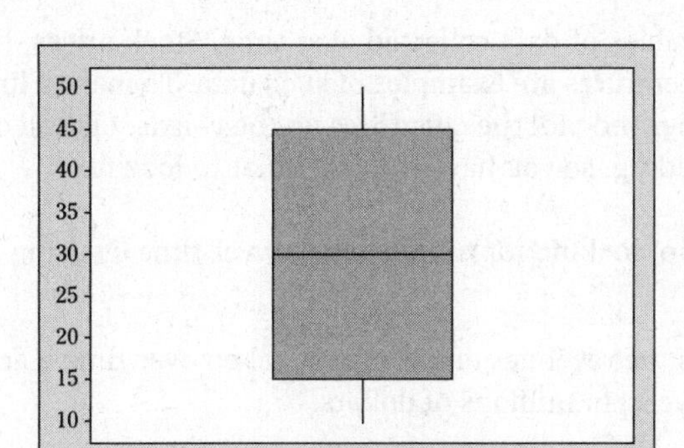

A. The shape of this box plot is symmetric because the lower part of the box is the same size as the upper part of the box. The center is the median (30), and the spread is indicated by the IQR, which is the third quartile (45) minus the first quartile (15), which equals 30. This is the range between the middle 50 percent of the values in the data set. The max and min are found to be 50 and 10, respectively.

19 Interpret the shape, center and spread of the following box plot.

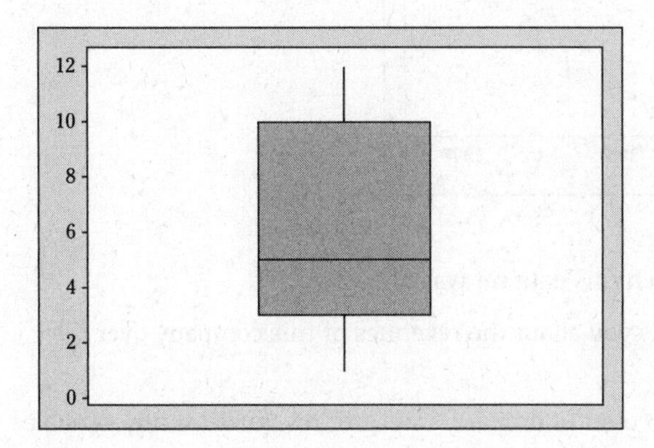

20 Interpret the shape, center and spread of the following box plot.

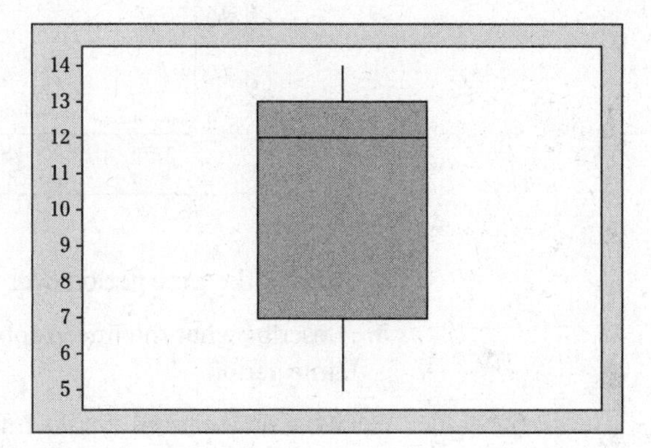

Looking at Line Graphs

A *line graph* shows quantitative values of data collected over time. Stock prices, the minimum wage, population sizes, and temperatures are examples of such data. To make a line graph, you simply order your data across time, and plot the quantities on the *y*-axis. Like all other types of graphs, line graphs can be misleading, so you have to know what to look for.

See the following for an example of looking for trends in data over time by using a line graph.

EXAMPLE

Q. The following figure shows the revenues of a company taken over time. Each dot represents the revenue for that year, in millions of dollars.

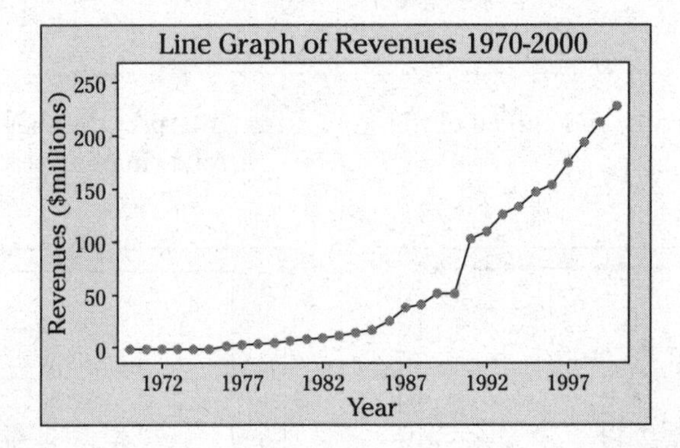

a. What's the time period over which this data set was collected?

b. Describe what the line graph tells you about the revenues of this company over the time period.

c. What do you need to take into account to properly interpret revenues (or any variable reported in dollars) over time?

A. Be aware of the impact of inflation over time on data reported in dollars. Some line graphs adjust for inflation and some don't. Look at the fine print.

a. The time period is approximately 1970–2000.

b. Revenues increased a little in the 1970s and then began a more steady increase in the '80s until around 1989, when the company broke the $50 million barrier. The company experienced a big jump around 1990–1991, with a very strong and steady increase each year since. In 2000, the company's revenue was up to $225 million and rising.

c. You should take inflation of the dollar over time into account. The revenues may look larger later on, but the value of the dollar has decreased over time as well. Some line graphs adjust for inflation.

21 Check out the sales of a particular car across the United States over a 60-day period in the following figure.

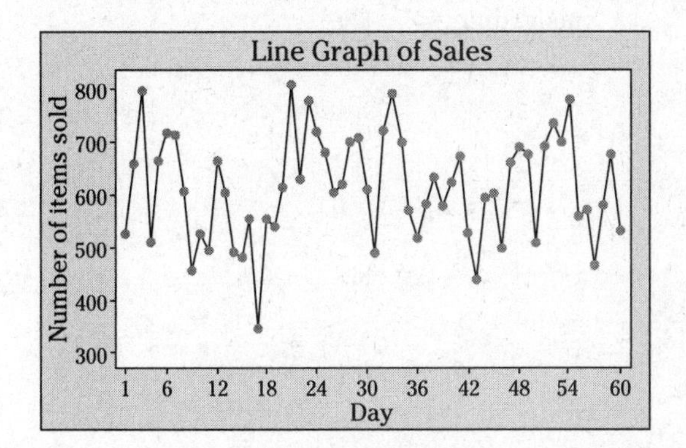

a. Can you see a pattern to the sales of this car across this time period?

b. What are the highest and lowest numbers of sales, and when did they occur?

c. Can you estimate the average of all sales over this time period?

22 Bob decides that after his heart attack is a good time to get in shape, so he starts exercising each day and plans to increase his exercise time as he goes along. Look at the two line graphs shown in the following figures. One is a good representation of his data, and the other should get as much use as Bob's treadmill before his heart attack.

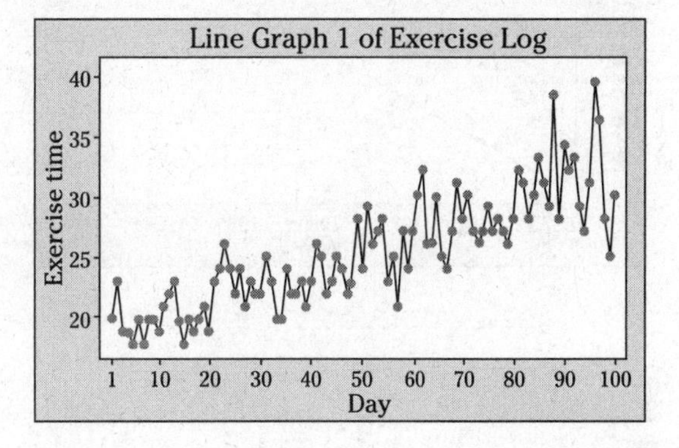

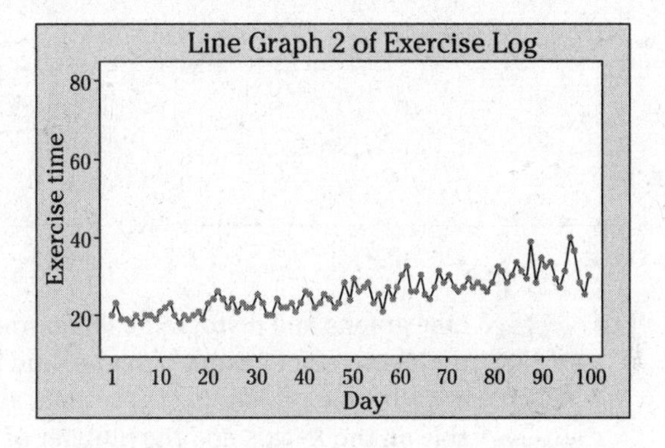

a. Compare the two graphs. Do they represent the same data set, or do they show totally different data sets?

b. Assume that both graphs are made from the same data. Which graph is more appropriate and why?

23 The line graph in the following figure shows one company's revenues over time. Explain why this graph is misleading and what you can do to fix the problem.

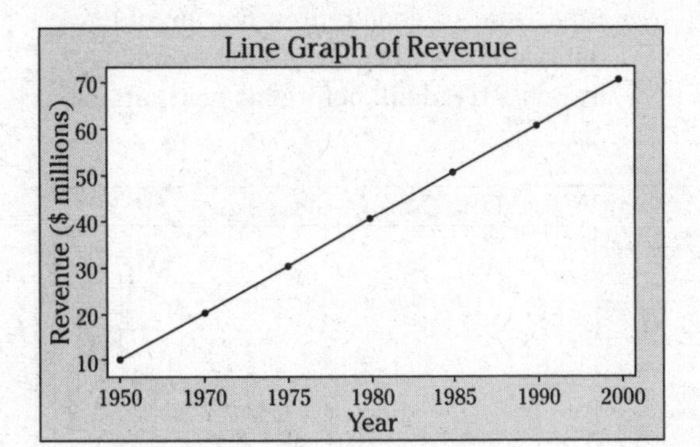

Line graphs typically connect the dots that represent the data values over time. If the time increments between the dots are large, explain why the line graph can be somewhat misleading.

WARNING

Line graphs and histograms do not measure the same things. Line graphs have the variable on the Y-axis (for example revenue) and time on the X-axis (for example year). It shows one point at each time period that represents all the data at that time period. A histogram has the variable on the X-axis and the number of individuals with that value of the variable on the Y-axis. It only measures one variable, not two, like a line graph does. And, it shows all the data at that one moment in time, stretching out over all the bars in the graph. You could think of each point on a line graph as representing an entire histogram of data if you will.

Understanding the Empirical Rule

The *empirical rule*, also known as the 68-95-99.7 rule, helps you pull together the ideas of mean and standard deviation for data sets that have a mound shape. A *mound-shaped* data set is symmetric, with one big mound (or mode) in the middle of the data set, sloping down on either side at the same rate.

A mound-shaped data set has certain characteristics, noted by the empirical rule. If you start at the mean and move one standard deviation on either side, you account for about 68 percent of the data because most of the data are in this middle, or mound, area. If you move another standard deviation away from the mean on either side (a total of two standard deviations from the mean), you account for about 95 percent of the data. You've covered the whole mound area, and most of the tail area as well, by this point. To cover most of the remaining data in the last part of the tails, you go out a third standard deviation on either side of the mean.

See the following for an example of checking before using the empirical rule.

EXAMPLE

Q. Does the empirical rule apply to the data set in the following figure? Explain.

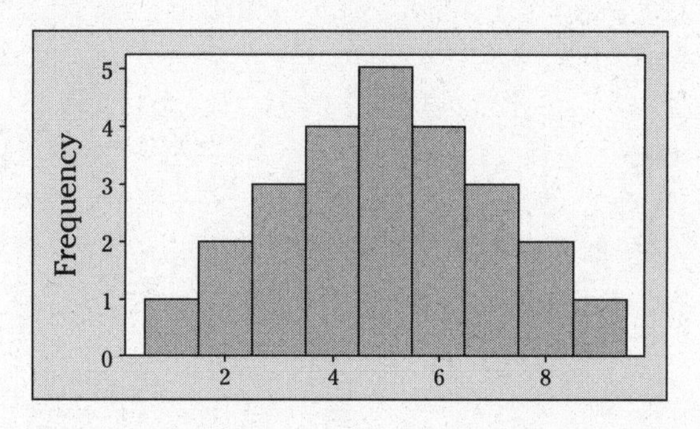

A. Yes, the empirical rule applies because the data set is mound-shaped.

---

25 Does the empirical rule apply to the data set shown in the following figure? Explain.

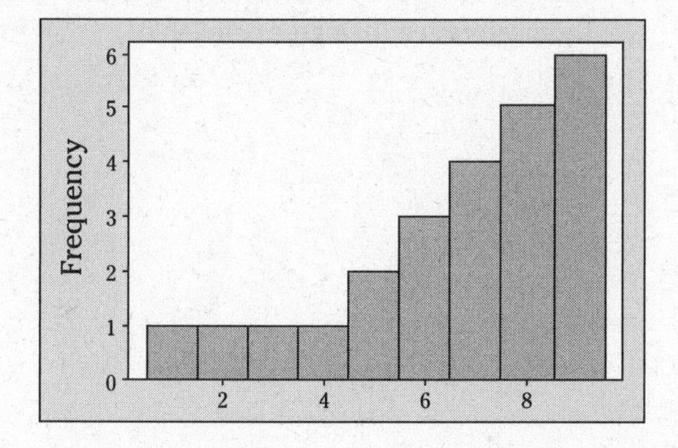

26 Suppose that a driver's test has a mean score of 7 (out of 10 points) and standard deviation 0.5.

a. Explain why you can reasonably assume that the data set of the test scores is mound-shaped.

b. For the drivers taking this particular test, where should 68 percent of them score?

c. Where should 95 percent of them score?

d. Where should 99.7 percent of them score?

27 Suppose that you have a data set of 1, 2, 2, 3, 3, 3, 4, 4, 5, and you assume that this sample represents a population. The mean is 3 and the standard deviation is 1.225.

a. Explain why you can apply the empirical rule to this data set.

b. Where would "most of the values" in the population fall, based on this data set?

28 Suppose that a mound-shaped data set has a mean of 10 and standard deviation of 2.

a. About what percentage of the data should lie between 8 and 12?

b. About what percentage of the data should lie above 10?

c. About what percentage of the data should lie above 12?

Suppose that a mound-shaped data set has a mean of 10 and standard deviation of 2.

a. About what percentage of the data should lie between 6 and 12?

b. About what percentage of the data should lie between 4 and 6?

c. About what percentage of the data should lie below 4?

Explain how you can use the empirical rule to find out whether a data set is mound-shaped, using only the values of the data themselves (no histogram available).

Answers to Problems in Organizing Quantitative Data

1. You don't know the actual values of the data anymore; you know only what group they fall into. For example, if eight test scores fall in the group 70–79, they could all be 70, they could all be 79, or they could be some mixture in between.

2. One possible histogram of this data is shown in the following figure. Yours may have different groupings and look slightly different. The data is quantitative, so a pie chart isn't appropriate. The groupings I chose for this histogram are as follows: 48–52; 53–57; 58–62; 63–67; 68–72; 73–77; 78–82; 83–87; 88–92; 93–97; and 98 and up.

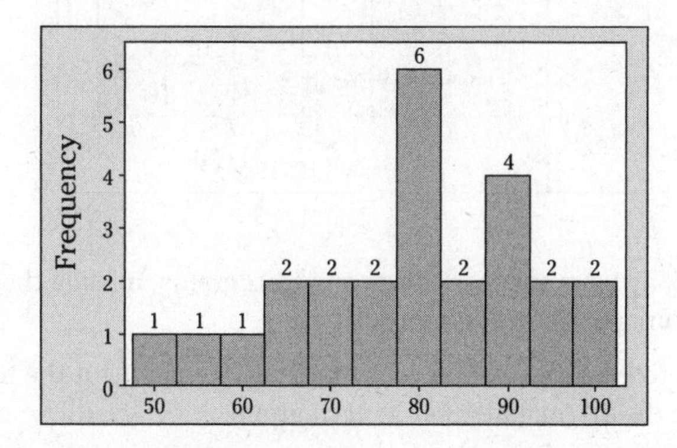

TIP

The scale of the X-axis is continuous on a histogram, so if you have a place for a bar but have no data, you should leave a space in that spot. If you don't leave a space, you'll have an incorrect histogram, because the gaps help you to distinguish how spread out the data is throughout the bars.

3. The following figure shows the relative frequency histogram for this data. Yours should look similar. The relative frequencies are $2 \div 45 = 0.044$ or 4.4%; $17 \div 45 = 0.38$ or 38%; $22 \div 45 = .49$ or 49%; $3 \div 45 = 0.07$ or 7%; and $1 \div 45 = 0.02$ or 2%.

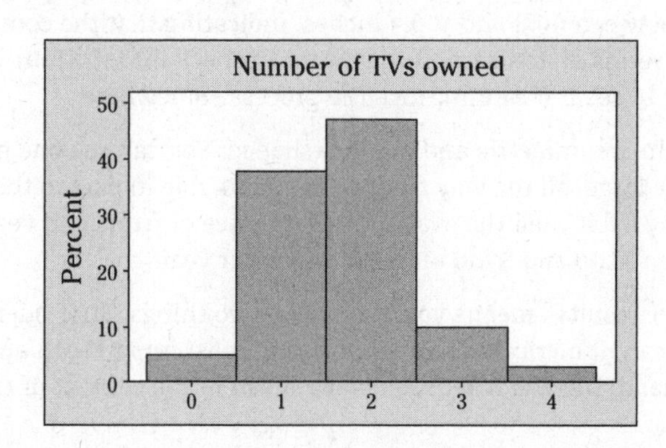

4. You can find the total sample size from a frequency histogram.

   a. The relative frequency histogram is shown in the following figure. *Note:* The total sample size is missing here, but you can still find it by summing up the heights of all the bars. Here, the total frequency is $10+15+5+15+5+20=70$. To get the relative frequencies, divide the height of each bar by 70 (the total sample size) to obtain a percentage. For example, the percentage of ones is $10 \div 70 = 14.3\%$, the height of the first bar.

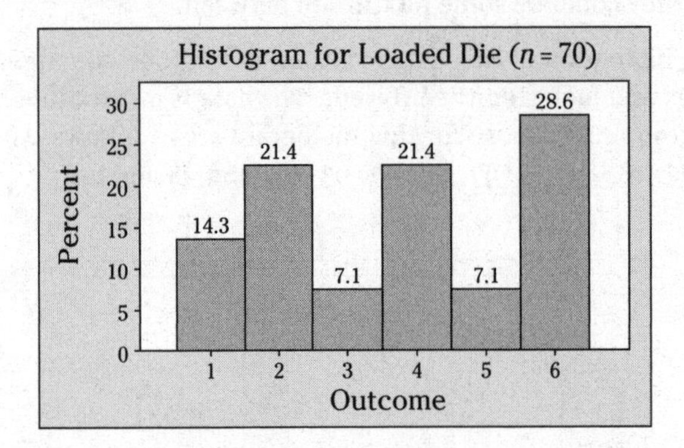

TIP

When making your own bar charts and histograms, include the total sample size if you want to score points with your professor.

   b. No. If you receive only the percent in each group, without the total sample size, you can't determine the original number in each group.

5. The shape of this data set is skewed right. The median is somewhere between $150 and $200, and the mean is slightly larger because of the single $300, $400, and $500 withdrawals. A crude measure of spread without these three single values is less than $50 (taking the range without those values divided by 6); with these three values, a crude measure of spread is around $80 (the range of all the values divided by 6). You can see a more accurate measure of spread in Chapter 2.

6. The histogram of rod lengths is skewed left. It appears to be centered very near 100, but the mean drops down a bit because of the one value lower than the rest (99.2). The range of the lengths is very tight, between 100.6 and 99.2 inches (each rod is within 1.4 inches). Most rod lengths are between 99.8 and 100.3 inches, indicating that the company seems to be doing very well, in terms of accuracy. It may want to check into the four values of 100.6 and the one value of 99.2 to see if it can improve the process somewhere.

7. The histogram is symmetric and mound-shaped. You can see one peak around $500. The range is quite large, all the way from $100–$800, due in part to the size of the household and also to dietary habits and the availability and price of fruits and vegetables. Most households spent between $200 and $700 on these items per year.

REMEMBER

"Interpret the results" means you need to do two things. First, identify the statistics and what they mean numerically, and second (and most important), apply those statistics to the scenario at hand. Discuss what the results mean in the context of the problem.

8. *Percentage return* means the difference between the ending value minus the beginning value, divided by the beginning value; it can be a positive or negative number or zero for no change.

   a. The *x*-axis represents the recorded variable, which is the one-year return for each stock. A negative number means that the stock lost money during that year (the price at the beginning was more than its value at the end).

   b. The *x*-axis will contain negative values if some of the data are negative. However, the *y*-axis of a histogram reports the counts or percentage of data in each grouping, and those values are always greater than or equal to zero.

9. Be aware that people may be tempted to give an incorrect response to a salary question (in other words, lie). Such a fib is called *response bias,* and it results in data that's systematically over or under the truth (in this case, probably over).

   a. The graph shows that this group is making plenty of money, although the graph is skewed to the right, with fewer graduates making the large amounts. The range is quite large, going from $50,000–$100,000, and the center is probably in the high $60,000 range.

   b. Because the heights of the bars sum to 13 $(5+3+1+2+1+1)$, you know 13 salaries make up the data set. The median is the one in the middle, which is the 7th salary. Because the first bar contains five salaries and the second bar contains three, the 7th number is in the second bar, which is in the $60,000 range.

   c. You have only 13 salaries — from a "large" program — so the data probably isn't precise, because it represents such a small part of the overall group. Also, you need to keep in mind that these values will change over time.

10. The first figure in Question 10 has a smaller spread because the bars are higher in the middle, near the mean. Therefore, on average, many of the values in this data set are close to the mean. The second figure shows all the data spread out about equally, which indicates that some are close to the mean, but an equal percentage are a medium distance away, and another equal percentage are a long distance away. This results in a larger spread.

WARNING

Don't interpret a flat histogram as meaning that the data shows "no change" or no spread. A flat histogram means the data does have quite a bit of spread. Data with a bell shape that's tight around the middle has a much smaller spread, because you generally measure spread as average distance from the middle.

TIP

Note that the values on the *x*-axis are the same for both graphs. If the values aren't the same, you'll have difficulty comparing the spreads alone. You should also take the scale into account. This may be above and beyond your particular course, but if you want to compare spreads with data sets that have different scales and different means, you can use the *coefficient of variation,* which is the standard deviation divided by the mean. A large coefficient of variation means that the spread is large, relative to the mean. A small coefficient of variation means that the spread is small, relative to the mean.

11. Just knowing the mean and median can tell you a lot about a data set.

    a. This data set is skewed to the right. A few people in the company have very large salaries compared to the others, driving the mean up and away from most of the data, yet the median remains unaffected by the skew, staying in the "middle" of the data.

    b. The median is the most appropriate measure of center to use when the data set is skewed because outliers don't affect it.

    c. The employees would want to report the median salary because it's lower and better represents most employees in the company. The employers may try to use the mean salary, because it's higher and indicates what they actually have to pay overall for their employees.

TIP

If the data are skewed right, the mean is higher than the median. Conversely, if the data are skewed left, the mean is lower than the median. If the data are symmetric, the mean and median are the same (or very close).

12. Because the data set is skewed left, you have a few values that are lower than the rest, which drives down the mean. So the mean must be 19, and the median must be 38.

TIP

This is another point that instructors hammer home: the relationship between the mean and median for skewed data. Remember that skewed right means a few large values, and skewed left means a few small values. This helps you think about how the mean and median compare.

13. Where the mean shows up among the data in a data set depends on the shape of the data. The mean can be higher than most of the values; in this case, the data would be skewed to the right. The median can't be higher than most of the values; it's higher than 50 percent of the values and lower than 50 percent of the values, putting it right in the middle.

14. The standard deviation is based on the average distance from the mean. Outliers (which often appear in skewed data) influence the mean and, therefore, influence the standard deviation. Outliers drive up the standard deviation, making the average distance from the mean seem larger than it is for most of the data. How do you get around this? You can report the *interquartile range*, the difference between the 75th and 25th percentiles (Earlier in this chapter we refer the reader to Chapter 2).

TIP

How can you tell whether a point in the data set is an official outlier? Statisticians have a few general rules but nothing really hard and fast. You can look at the histogram and calculate your statistics with and without the outliers to see how the numbers change. The obvious outliers are the ones that change the statistics the most.

15. No, the die isn't loaded; the graph is loaded. The histogram is misleading because it starts at 40 on the *y*-axis and goes to only 65. The differences in the heights of the bars are exaggerated because the graph doesn't start at zero. A more fair and balanced-looking graph of the same data is shown here. Notice the frequencies for the outcomes are fairly close, indicating no evidence of the die being loaded.

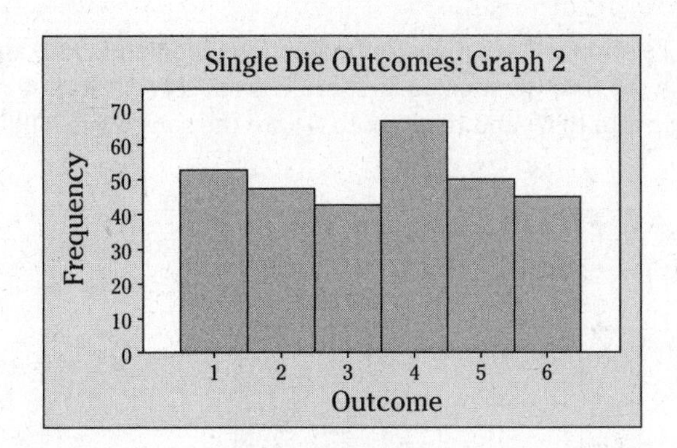

Beware of graphs that don't start at zero on the y-axis. They may make the results look more dramatic than necessary, which is misleading to the reader.

(16) This histogram has a great deal of open space at the top, and the bars appear to be very short and close together in height. The scale for the y-axis of this graph uses oversized increments. A better histogram would have smaller increments on the y-axis, and it wouldn't include numbers that go beyond what you need to show the data. Such a histogram follows.

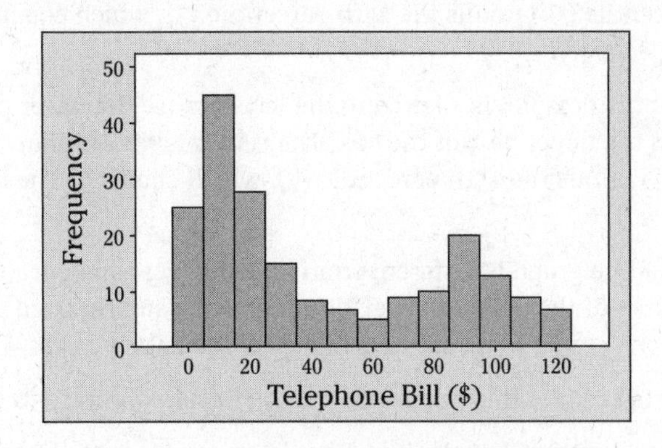

(17) Notice that this graph is skewed to the right because there is more space in the top area of the graph than the bottom area of the graph. The min and max are at 3 and 17, respectively, and the IQR shows the middle 50% of the data has a range of $16 - 4 = 12$.

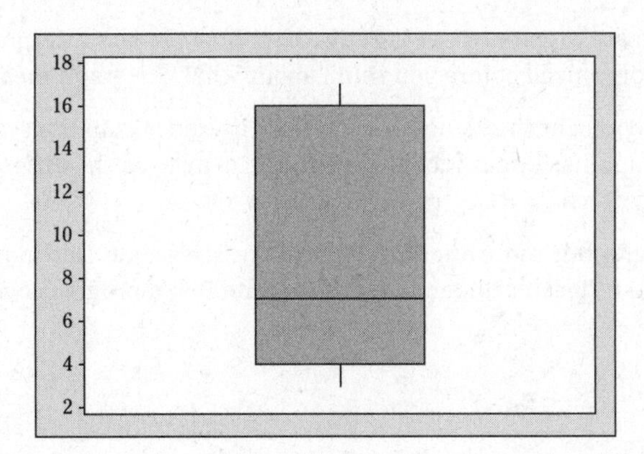

18 This graph is symmetric right down the line. The median is 120, and there is the same distance from Q1 (105) to the median as there is from Q3 (135) to the median. In addition, the lines from the min to Q1 and the max to Q3 are the same. The IQR is $135 - 105 = 30$.

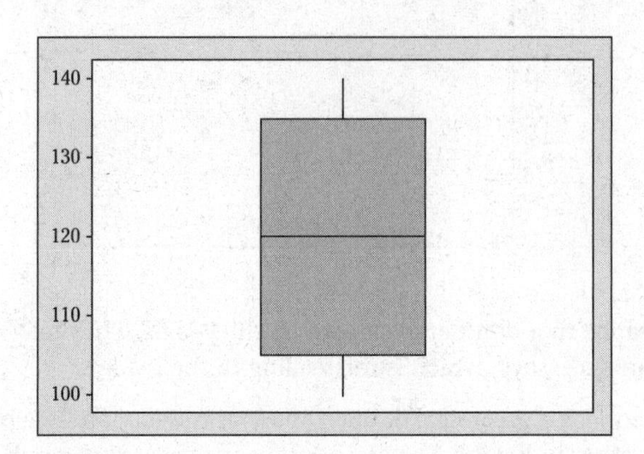

19 The shape of this box plot is skewed to the right because the upper part of the box covers more distance than the lower part of the box. The center is the median (5), and the IQR is the 75th percentile (10) minus the 25th percentile (3), which equals 7. The min and max are 1 and 12, respectively.

20 The shape of this box plot is skewed to the left because the lower part of the box covers more distance than the upper part of the box. The center is the median (12), and the IQR is the 75th percentile (13) minus the 25th percentile (7), which equals 6. The min and max are 5 and 14, respectively.

21 Interpreting a line graph is different from interpreting a histogram. A line graph represents many snapshots of the situation over time, each one summarized by one point, and a histogram shows one single snapshot in detail of all the data at once.

a. No, the data seem to fluctuate back and forth from around 350 to 800 cars sold per day.

b. You can't tell exactly, but the highest sales figure is around 800, coming on days 3 and 21, and the lowest (350) occurred only a few days before, around day 17. Maybe the customers knew a sale was coming and waited to buy.

c. The average appears to be around 600 cars sold per day, looking at what the values on the *y*-axis seem to center around. (The actual average is 613.)

22 The way data are organized can greatly affect how readers interpret graphs. Look at the way the data are organized before you think about what the graph means.

a. They do represent the same data set; the difference is in the scale used on the *y*-axis. The second graph has larger increments on the *y*-axis, so the differences in the data over time are played down.

b. The first graph is more appropriate. It has a scale that uses most of the space on the graph, and it doesn't understate or overstate Bob's progress over time.

(23) This line graph is misleading because the time increments on the x-axis (time) aren't equally spaced, but the graph presents them as if they are, which incorrectly makes it look like the revenue is increasing at the same rate over time. To fix the problem, you need to space the time periods shown on the x-axis properly. The correct line graph is shown in the following figure.

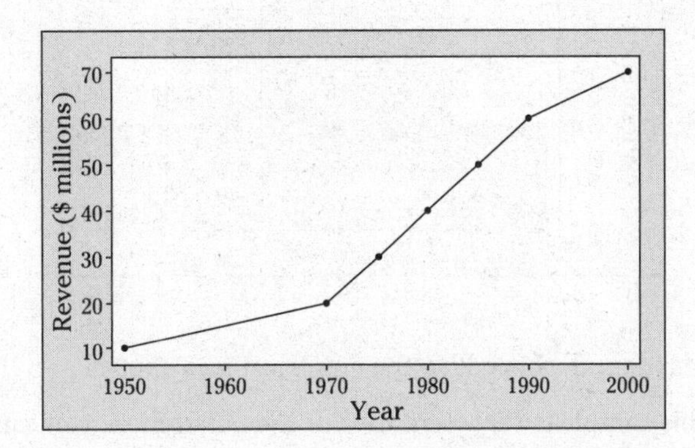

When you see a line graph, look at the time increments on the x-axis and make sure the times are equally spaced in terms of the number of years between them.

(24) If the time increments between the dots are large, and you connect the dots, you assume that the change that took place during the interim period (when data wasn't collected) occurred at a steady rate, represented by the line that connects the dots. That may not always be the case.

(25) No, because the data is skewed, it isn't mound-shaped. It has one mode, but it isn't symmetric.

REMEMBER

The empirical rule works only if the data is mound-shaped. It doesn't apply to skewed data. Check the shape of the data first before you attempt to apply the empirical rule. Don't get caught applying something that doesn't fit the data. (If you learned Chebyshev's theorem, use that for data that isn't mound- or bell-shaped.)

(26) Sometimes, you have to make reasonable assumptions before you proceed to solve a problem.

a. Test scores from a properly written and calibrated test are likely to be mound-shaped, with equal numbers of people scoring below the mean as scoring above the mean.

b. The empirical rule says that about 68 percent of the test-takers should score within one standard deviation of the mean. In other words, $7 \pm (1 * 0.5) = 7 \pm 1(0.5)$. The lower limit is $7 - 0.5$, which is 6.5, and the upper limit is $7 + 0.5$, which is 7.5.

c. The empirical rule says that about 95 percent of test-takers should score within two standard deviations of the mean. In other words, $7 \pm 2(0.5)$. The lower limit is $7 - 1$, which is 6, and the upper limit is $7 + 1$, which is 8.

d. The empirical rule says that about 99.7 percent of test-takers should score within three standard deviations of the mean. In other words, $7 \pm 3(0.5)$. The lower limit is $7 - 1.5$, which is 5.5, and the upper limit is $7 + 1.5$, which is 8.5.

(27) The histogram for this data is shown in the following figure.

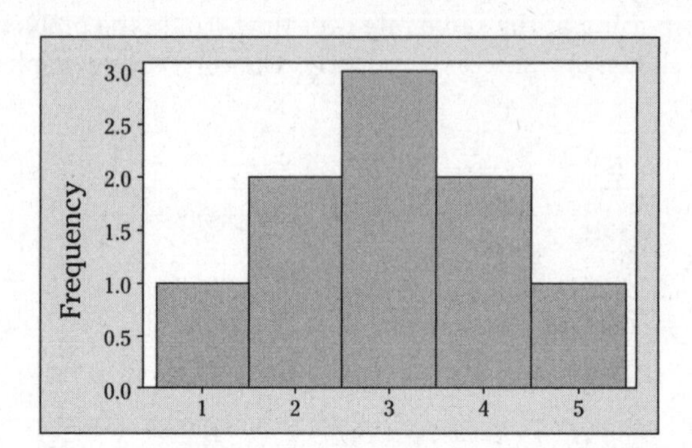

a. This data is mound-shaped, so the empirical rule applies.

b. Because this sample is representative of the population, you can say that most of the values in the population should lie within two standard deviations of the mean of the sample. The mean is 3, and the standard deviation is 1.225. This indicates that 95 percent of the values in the population should lie between $3-2*1.225$ and $3+2*1.225$; in other words, between 0.55 and 5.45.

TIP

"Most of the data" can mean 68 percent, 95 percent, or 99.7 percent, but most statisticians would say that 95 percent is the magic number when you want to discuss where "most" of the data lies. Be sure you define what you mean by "most" if you're asked to describe where "most" of the data is.

(28) This problem combines the empirical rule with the idea that the total of all the relative frequencies of a data set has to equal 1.

a. Because 8 and 12 are both one standard deviation from the mean, the percentage of data lying between them according to the empirical rule is about 68 percent.

b. Because mound-shaped data are symmetric, half of the data should lie above the mean and half of the data should lie below the mean. So the answer is 50 percent.

c. Because about 68 percent of the data lie between 8 and 12, $68 \div 2 = 34\%$ of the data lie between 8 and 10, and the other 34% lie between 10 and 12. Because half of the data lie above 10, and 34% of the data lie between 10 and 12, $50\% - 34\% = 16\%$ of the data should lie above 12. Take what you know and subtract off the part you don't need to get the part that you want. See the following figure for an illustration of these ideas.

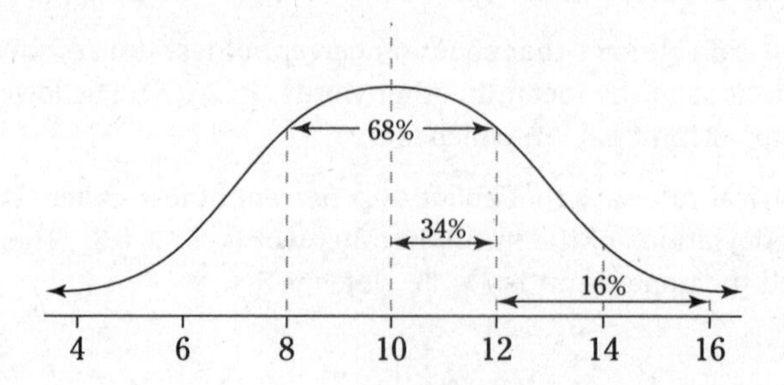

A picture may be very helpful for problems involving the empirical rule. It helps you to stay organized in your thinking, and it helps your professor see what you're doing, in case you make a mistake in your calculations somewhere along the line. Mark off the mean and the standard deviations to the left and right (go out three of them on each side), and then shade in the area for which you want to find the percentage.

(29) This problem uses different parts of the empirical rule in tandem.

a. The number 6 is two standard deviations below the number 10 because $10-2(2)=6$. You know by the empirical rule that about 95% of the data set lies between 6 and 14, so half of that (or 47.5%) lies between 6 and 10 by symmetry. That gives you one piece of what you need. Now you need the part from 10 to 12. Twelve is one standard deviation away from the mean, and you know that about 68% of the values lie between 8 and 12, so 34% of the data lie between 10 and 12. This gives you the other piece that you need. Add the two pieces together to get $47.5\%+34\%=81.5\%$ of the data between 6 and 12.

When using the empirical rule, add the two pieces together when you want the percentage of data that fall between the two numbers when one is below the mean and the other is above the mean.

b. This time you find the two pieces and subtract them, because you want the area between them. Because 6 is two standard deviations below 10 since $10-2(2)=6$, you know that about $95 \div 2=47.5\%$ of the data lie between 6 and 10. Because 4 is three standard deviations below 10 since $10-3(2)=4$, about 99.7% of the data lie between 4 and 16. That means half of the data, or 49.85%, should lie between 4 and 10. You want the area between 4 and 6, so take the area between 4 and 10 (49.85%) and subtract the area you don't want, which is the area between 6 and 10 (47.5%). You get $49.85\%-47.5\%=2.35\%$.

Percents are never negative, so make sure you take the bigger area first and subtract off the smaller area that you don't want to find the percentage of the data that fall between two numbers. This subtracting is done only when both numbers are on the same side of the mean and the empirical rule is being used.

c. To get the area below 4, take the area from 4 to 10, which is about $99.7\% \div 2\%=49.85\%$, and subtract that from 50% (0.5) because half of the data lies below 10 by symmetry. So the answer is about $50\%-49.85\%=0.15\%$ of the data should lie below 4.

(30) You can calculate the mean plus or minus one standard deviation to get the upper and lower limits. According to the empirical rule, if the data are mound-shaped, about 68 percent of the data should lie between these two values. Count how many data points are actually in this interval and divide by the sample size. If this percentage isn't close to 68, the data isn't mound-shaped. If it is, move on to the next standard deviation. Take the mean plus or minus two standard deviations and find those limits. Find the percentage of data lying between those two numbers and compare it to 95 percent. If the figure is close, go on to the third step (99.7 percent). All three criteria have to be met to say that the data are mound-shaped.

2

Probability, Distributions, and the Central Limit Theorem (Are You Having Fun Yet?)

IN THIS PART . . .

Understand the basic underpinnings of statistics — distributions.

Tackle sampling distributions and margin of error.

Become familiar with standard scores, probability, and percentiles from the Z-distribution.

Get the ins and outs of the t- and binomial distributions.

» Steering clear of probability misconceptions

» Forecasting the future with probabilities

Chapter 5

Understanding Probability Basics

In this chapter, you practice using and applying the rules and ideas of probability. The rules and laws of probability often go against our intuition. For example, a combination like 1, 2, 3, 4, 5, 6 seems a very unlikely winner in the lottery, when, in fact, it has just the same chance as any other combination (which shows how *un*likely they all really are). And it might seem that a couple that has three girls in a row should have a much higher chance of having a boy this time, but barring any genetic predisposition to girls, the chance is still 50-50 each time. The practice in interpreting probability that this chapter provides helps ease your statistical state of mind and allows you to avoid making some of the more common misconceptions about the subject.

Grasping the Rules of Probability

The definition of the *probability* of an outcome is the long-run percentage (or relative frequency) of times that you expect the outcome to occur. Probabilities follow certain rules. Here they are in a nutshell:

» Every probability is a percentage between 0 and 100. In terms of decimals, a probability (proportion) is a number between 0 and 1. A probability of 1 means the outcome is certain, and a probability of 0 means the outcome is impossible.

» The set containing all possible outcomes is called the *sample space*. The sum of the probabilities of all possible outcomes in the sample space is equal to 1.

» The probability of an event or set of outcomes that are *disjoint* (sharing no outcomes in common) is equal to the sum of the individual probabilities for each disjoint outcome.

» The complement of an event is all the possible outcomes except those that make up the event. The probability of the complement of an event is 1 minus the probability of the event.

The way to understand probability and how it works is to start with a small example, such as the following, and work from there.

EXAMPLE

Q. Suppose that you flip a fair coin three times.

a. How many outcomes are possible?

b. What's the probability of each outcome?

c. What are the possible values for the total number of heads out of three tosses, and what are their probabilities?

A. Before you start flipping your coins, it's good to know exactly what the possibilities and their probabilities are. The table at the end of this example shows you the possibilities and probabilities at a glance for this coin-flipping scenario.

a. The sample space contains eight outcomes: HHH, HHT, HTH, THH, HTT, THT, TTH, TTT. Notice that each flip has two possible outcomes, so three flips has $2*2*2=8$ possible outcomes.

b. Each outcome has a probability of 1 in 8 (because the total number of outcomes is eight, and because you assume the coin to be two-sided and fair).

c. The total number of heads can be 3, 2, 1, or 0. Three heads happens only one way, HHH, so its probability is 1 in 8. Two heads happens three ways, HHT, HTH, THH, so its probability is 3 in 8. One head happens three ways, HTT, THT, TTH, so its probability is 3 in 8. Zero heads happens in only one way, TTT, so its probability is 1 in 8. All these probabilities sum to 1.

Number of Heads	Possible Outcomes	Probability
3	HHH	1 in 8
2	HHT, HTH, THH	3 in 8
1	HTT, THT, TTH	3 in 8
0	TTT	1 in 8

1 M&Ms colors come in the following percentages: 13 percent brown, 14 percent yellow, 13 percent red, 24 percent blue, 20 percent orange, and 16 percent green. Reach into a bag of M&Ms without looking.

a. What's the chance that you pull out a brown or yellow M&M?

b. What's the chance that you won't pull out a blue?

2 Suppose that you flip a coin four times, and it comes up heads each time. Does this outcome give you reason to believe that the coin isn't legitimate?

3 Consider tossing a fair coin 10 times and recording the number of heads that occur.

a. How many possible outcomes would occur?

b. What would be the probability of each of the outcomes?

c. How many of the outcomes would have 1 head? What is the probability of 1 head in 10 flips?

d. How many of the outcomes would have 0 heads? What is the probability of 0 heads in 10 flips?

e. What's the probability of getting 1 head or less on 10 flips of a fair coin?

4 What's the probability of getting more than 1 head on 10 flips of a fair coin?

Avoiding Probability Misconceptions

Probability often goes against our intuition or our desires; we often ignore it or are unaware of its impact. For example, thinking that every situation with two possible outcomes is a 50-50 situation can get you into trouble (yet it seems tempting). And the biggest misconceptions are in the gambling arena, with people thinking they'll hit it big any minute now. The casinos are counting on you having that attitude (and are literally counting your money).

WARNING

Here are some common misconceptions about probability that you want to avoid:

- Believing that outcomes that appear to be "more random" have a higher chance of occurring than outcomes that don't
- Thinking probability works well for predicting short-run behavior
- Claiming you can be "on a roll" or "due for a hit" by the law of averages
- Treating any situation with only two possible outcomes as a "50-50" situation
- Misinterpreting a rare event

The problem with these misconceptions is that they appear to make sense, and you may want them to be true, but they just aren't. With probability practice, you can begin to see through the misconceptions and find out how to avoid them. (In the solutions at the end of this chapter, I elaborate more on these misconceptions as they come up in the practice problems. You can also find more discussion on probability misconceptions in *Statistics For Dummies*, 2nd Edition [Wiley], if you happen to have a copy.)

5 Suppose that an NBA player's free throw shooting percentage is 70 percent.

a. Explain what this means as a probability.

b. What's wrong with thinking that his chances of making his next free throw are 50-50 (because he either makes it or he doesn't)?

6 Suppose that you buy a lottery ticket, and you have to pick six numbers from 1 through 50 (repetitions allowed). Which combination is more likely to win: 13, 48, 17, 22, 6, 39 or 1, 2, 3, 4, 5, 6?

7. You feel lucky again and buy a handful of instant lottery tickets. The last three tickets you open each win a dollar. Should you buy another ticket because you're "on a roll"?

8. Suppose that a small town has five people with a rare form of cancer. Does this automatically mean a huge problem exists that needs to be addressed?

Making Predictions Using Probability

One of the most important functions of probability used by researchers and the media is its usefulness in making predictions. Forecasts can be very useful, but you have to temper them with the understanding that probability is a long-term predictor, not a short-term fix. You also need to realize that building a model to make a prediction can be a very complicated business.

See the following example for a problem dealing with probability and predictions.

EXAMPLE

Q. Suppose you know that the chance of winning your money back on an instant lottery ticket is 1 in 10. Does that mean if you buy 10 lottery tickets, you know one of them will be an instant winner?

A. No. This misconception of probability is a popular one. Probability is the long-term percentage of instant winners, and it doesn't apply to a sample as small as 10. Results will vary from sample to sample.

9 A couple has conceived three girls so far with a fourth baby on the way. Do you predict the newborn will be a girl or a boy? Why?

10 Meteorologists use computer models to predict when and where a hurricane will hit shore. Suppose they predict that hurricane Stat has a 20 percent chance of hitting the East Coast.

a. On what info are the meteorologists basing this prediction?

b. Why is this prediction harder to make than your chance of getting a head on your next coin toss?

11 Bob has glued himself to a certain slot machine for four hours in a row now with his bucket of coins and a bad attitude. He doesn't want to leave because he feels the longer he plays, the better chance he has to win eventually. Is poor Bob right?

12 Which situation is more likely to produce exactly 50 percent heads: flipping a coin 10 times or flipping a coin 10,000 times?

Answers to Problems in Probability

1. Probabilities apply to individual selections as well as to long-term frequencies.

   a. Because 13 percent of all plain M&Ms are brown, the chance that the one M&M you pick is brown is 13 percent. For yellow, the chance is 14 percent. For brown or yellow, add the probabilities to get 27 percent.

   b. Because 24 percent of plain M&Ms are blue, $100\% - 24\% = 76\%$ aren't blue, by the complement rule.

2. You don't have enough data to determine the probability until you look at the long-run percentage of heads, and four flips isn't a long run of data collecting. But another way to look at this question is to find the probability of flipping four heads on four flips of a coin. Flipping a coin four times gives you $2^4 = 16$ possibilities. You can get four heads one way, so the chance of flipping four heads on four flips is $1 \div 16 = 0.06$ or 6%. It doesn't happen very often, but it does happen. You may be skeptical of the coin, but you should collect more data before you decide.

3. This problem seems daunting until you realize you have an easy way and a hard way to do this problem (and your professor is banking on you realizing the same thing!).

   a. Flipping a coin 10 times results in $2*2*2\ldots*2$ or $2^{10} = 1{,}024$ possible outcomes. (Don't try to list them all out!)

   b. Because the coin is fair, each outcome is equally likely and has a probability of 1 in 1,024.

   c. You have 10 ways to get a single head on 10 flips. One way is to have HTTTTTTTTT. But the head can come on the second toss, the third toss, or anywhere up to the tenth toss. The probability is, therefore, 10 in $1{,}024 = 0.01$.

   d. To flip no heads, you have to get all tails, and that happens only one way out of 1,024, which equals 0.001.

   e. Add the two probabilities from Question 3c and 3d to get 11 in $1{,}024 = 0.011$ for the probability of less than or equal to 1 head on 10 flips of a coin.

4. Solving this problem is much easier if you look at the complement. The complement of flipping more than one head is flipping less than or equal to one head (a number line can help show you this as well). In Question 3, you find the probability of less than or equal to one head to be 11 in 1,024. The probability of flipping greater than one head is, therefore, $1-(11 \div 1{,}024) = 0.989$.

5. The misconception that random situations having only two possible outcomes are 50-50 situations is a very common one.

   a. The 70 percent shooting clip means that in the long term, over many free throws, this player makes his shots 70 percent of the time, on average.

   b. The 50-50 argument breaks down because the two outcomes aren't equally likely. According to past data, this player hits 70 percent of his free throws and misses only 30 percent of the time. Using these numbers is similar to flipping an unfair coin. You have two sides, yes, but the two outcomes aren't equally likely. If the 50-50 argument worked, we should all buy lottery tickets, because either you win or you don't with a 50-50 chance, right? (In our dreams!).

REMEMBER

Just because you have two possible outcomes doesn't mean they each have a 50 percent chance of happening. You have to look at past data and determine what the weight is for each outcome, just like in every other situation.

6 Both outcomes are equally likely. Assuming that the lottery process is fair, every single combination of six numbers has an equally likely chance of being selected. The combo 1, 2, 3, 4, 5, 6 seems like it could never happen, but the probability shows you how unlikely any combination is to win. After all, with 50 numbers to choose from and 6 numbers to pick, you have millions of possibilities. But here's a tip: Go ahead and pick 1, 2, 3, 4, 5, 6. If you do win, you won't have to split the money with anyone, because no one else is picking that combination! (Until they read this, at least . . .)

7 No. Probability is a long-term percentage. What recently happened has no impact on what happens in the future. Suppose that 5 percent of all instant lottery tickets are winners. This figure tells you nothing about when those winners will come up.

REMEMBER

Probability predicts long-term behavior only; it can't guarantee any kind of short-term outcomes. Numbers may take a long time to "average out," and there's no such thing as being "on a roll" or "due for a hit." The law of averages idea applies only to the long term.

8 Not necessarily. The chances of this happening are quite small, but even if the probability is one in a million, you should expect it to happen, on average, once out of every million times. Over a period of years, in a very large country, a situation like this is bound to happen just by chance. The outbreak may point to something else, and the town should investigate, but it doesn't automatically mean that a problem exists.

9 The chance of having a boy this time is the same as it was with the previous births, 1 in 2. Probability has no memory of recent happenings and can't predict short-term behavior.

10 Computer models use a process called *simulation.* You put all your data in, make a mathematical model out of it, and repeatedly run the computer through the scenario to see what happens.

a. The 20 percent comes from the fact that 20 percent of the times the computers repeated the scenario the results pointed to the hurricane hitting the coast, and 80 percent of the times had it veering off into sea.

b. Computer models are based on a great deal of information, but many assumptions fill in the blanks. Some of these assumptions can be wrong, thus throwing the percentage predictions right out the window.

11 Not really. In the long run, Bob should expect to lose a small amount with a very high probability every time he pulls the handle on the machine. Yes, he may win big on one pull, but the chance is so tiny that, on average, he doesn't have enough time in his life to sit on that stool and wait to win. Who makes all the money? The person who owns the slot machine, that's who.

12 Flipping a coin 10,000 times has a higher chance of landing 50 percent heads (exactly), because when you flip it only 10 times, the results still vary so much that you can get results like three, four, five, six, or seven heads with a fairly high probability. But when you increase the sample size to 10,000, the relative frequency (percentage of heads observed) becomes closer and closer to the true probability you expect (in this case, 50 percent). This shows the real law of averages at work.

REMEMBER

The law of averages says that, in the long run, the percentage of occurrences of an event gets closer and closer to the true probability of the event.

backward

» Translating to a standard normal (Z) distribution

» Calculating and interpreting Z-scores

» Finding and understanding percentiles for the normal distribution

Chapter 6
Measures of Relative Standing and the Normal Distribution

This chapter helps you get comfortable with the *normal distribution* — the most popular distribution in introductory statistics. You practice converting to the standard normal (Z) distribution and interpreting exactly what the distribution means. I show you how to find and interpret percentiles for the normal distribution, and you practice on "backwards normal" problems that instructors like so much (the problems that give you a percent and ask you to find the cutoff point).

Mastering the Normal Distribution

You may have heard of a *bell curve*. A bell curve describes data from a variable that has an infinite (or very large) number of possible values distributed among the population in a bell shape. This basically means a big group of individuals gravitate near the middle, with fewer and fewer individuals trailing off as you move away from the middle in either direction. Statisticians call a distribution with a bell-shaped curve a *normal distribution.* You can see a normal distribution's shape in Figure 6-1.

FIGURE 6-1: Laypeople call it a bell curve; you call it a normal distribution.

Every normal distribution has certain properties. You can use these properties to determine the relative standing of any particular result on the distribution. The properties of any normal distribution (bell curve) are as follows:

- The shape is symmetric.
- The distribution has a mound in the middle, with tails going down to the left and right.
- The mean is directly in the middle of the distribution. (The mean of the population is designated by the Greek letter μ.)
- The mean and the median are the same value because of the symmetry.
- The standard deviation is the distance from the center to the *saddle point* (the place where the curve changes from an "upside-down-bowl" shape to a "right-side-up-bowl" shape. (The standard deviation of the population is designated by the Greek letter σ.)
- About 68 percent of the values lie within one standard deviation of the mean, about 95 percent lie within two standard deviations, and most of the values (99.7 percent or more) lie within three standard deviations by the empirical rule (see Chapter 4). (You practice on the tables in this chapter to get more specific.)
- Each normal distribution has a different mean and standard deviation that make it look a little different from the rest, yet they all have the same bell shape.

See the following for an example of comparing two normal distributions.

EXAMPLE

Q. Which of these two normal distributions has a larger standard deviation?

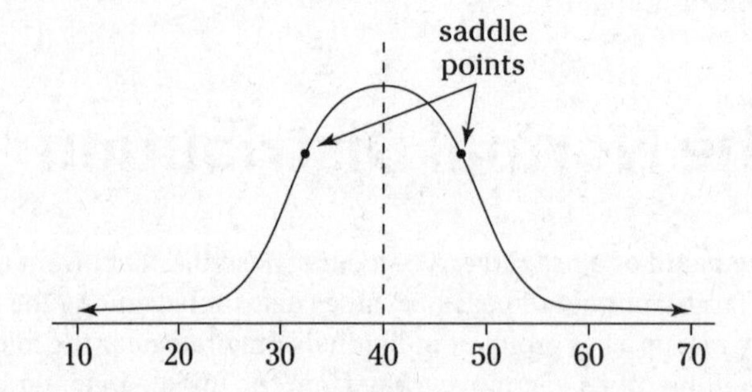

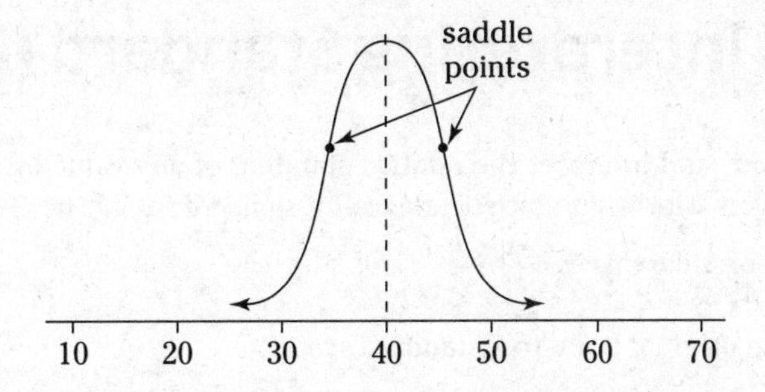

A. The first distribution has a larger standard deviation, because the data are more spread out from the center, and the tails take longer to go down and away.

1 What do you guess are the standard deviations of the two distributions in the previous example problem?

2 Draw a picture of a normal distribution with mean 70 and standard deviation 5.

3 Draw one picture containing two normal distributions. Give each a mean of 70, one a standard deviation of 5, and the other a standard deviation of 10. How do the distributions differ?

4 Suppose that you have a normal distribution with mean 110 and standard deviation 15.

a. About what percentage of the values lie between 110 and 125?

b. About what percentage of the values lie between 95 and 140?

c. About what percentage of the values lie between 80 and 95?

Finding and Interpreting Standard (Z) Scores

To find, report, and interpret the relative standing of any value on a normal distribution, you need to convert it to what statisticians call a *standard score*. The Z-formula for standardizing your value looks like this: $z = \frac{x - \mu}{\sigma}$.

To convert an original score to a standard score:

1. **Find the mean and the standard deviation of the values you're working with.**
2. **Take the value you want to convert and subtract the mean.**
3. **Divide your result by the standard deviation.**

Standard scores, also known as *Z-scores,* have a universal interpretation, which is what makes them so great.

TIP

Instructors love to ask you to interpret standard scores. If an instructor gives you a standard score, be ready to interpret it right away. For example, a standard score of +2 on an exam says that the score is two standard deviations above the mean, which is quite good in this case. If you're measuring times it takes to run around the block, however, a standard score of +2 would be a bad thing, because your time was two standard deviations above the mean. In this case, a standard score of –2 would be much better, indicating your time was less than most of the other runners. To interpret a standard score, you don't need to know the original score, the mean, or the standard deviation. The standard score gives you the relative standing of a value, which, in most cases, is what matters most.

When your data has a normal distribution, its Z-scores have a special normal distribution — one that has mean 0 and standard deviation 1. This distribution is called the *standard normal distribution* or the *Z-distribution* (see Figure 6-2).

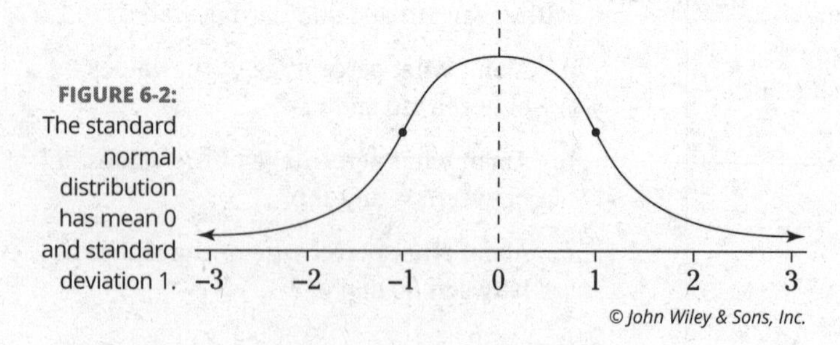

FIGURE 6-2: The standard normal distribution has mean 0 and standard deviation 1.

See the following for an example of interpreting a standard score.

EXAMPLE

Q. Suppose that you play a round of golf and want to compare your score to the other members of your club population. You find that your score is below the mean.

a. What does this tell you about your standard score?

b. Is this a good thing or a bad thing?

A. To interpret the standard score, the sign is the first part to look at.

a. Falling below the mean indicates a negative standard score.

b. Often, being below the mean is a bad thing, but in golf you have an advantage because golf scores measure the number of swings you need to get around the course, and you want to have a low golf score. So being below the mean in this case is a good thing.

5 Exam scores have a normal distribution with mean of 70 and standard deviation of 10. Bob's score is 80. Find and interpret his standard score.

6 Bob scores 80 on both his math exam (which has a mean of 70 and standard deviation of 10) and his English exam (which has a mean of 85 and standard deviation of 5). Find and interpret Bob's Z-scores on both exams to let him know which exam (if either) he did better on. Don't, however, let his parents know; let them think he's just as good at both subjects.

7 Sue's math class exam has a mean of 70 with a standard deviation of 5. Her standard score is –2. What's her original exam score?

8 Suppose that your score on an exam is directly at the mean. What's your standard score?

9. Suppose that the weights of cereal boxes have a normal distribution with a mean of 20 ounces and standard deviation of half an ounce. A box that has a standard score of 0 weighs how much?

10. Suppose that you want to put fat Fido on a weight-loss program. Before the program, his weight had a standard score of +2 compared to dogs of his breed/age, and after the program, his weight has a standard score of –2. His weight before the program was 150 pounds, and the standard deviation for the breed is 5 pounds.

    a. What's the mean weight for Fido's breed/age?

    b. What's his weight after the weight-loss program?

Knowing Where You Stand with Percentiles

Percentiles are another way to measure where you stand in a data set. If you come in at the 90th percentile, for example, 90 percent of the values are below you (and 10 percent are above you). In general, being at the *k*th percentile means *k* percent of the data lie below that point and $100 - k$ percent lie above it.

To calculate a percentile when the data has a normal distribution:

1. **Convert the original score to a standard score by taking the original score minus the mean and dividing by the standard deviation (in other words, use the Z-formula).**
2. **Use the Z-table (in the Appendix) to find the corresponding percentile for the standard score.**

   The values on the Z-table have a standard normal distribution (the special normal distribution with a mean of 0 and standard deviation of 1). The distribution is made up entirely of standard scores, or Z-scores, and is often called the Z-distribution.

See the following for an example of finding a percentile for a normal distribution.

EXAMPLE

Q. Weights for single dip ice-cream cones at Adrian's have a normal distribution with a mean of 8 ounces and standard deviation of one-quarter ounce. Suppose that your ice-cream cone weighs 8.5 ounces. What percentage of cones are smaller than yours?

A. Your cone weighs 8.5 ounces, and you want the corresponding percentile. Before you can use the Z-table to find that percentile, you need to standardize the 8.5 — in other words, run it through the Z-formula. This gives you $z = \frac{8.5 - 8}{.25} = \frac{.5}{.25} = +2$, so the Z-score for your ice-cream cone is +2. Now you use the table and find that the +2.0 in the standard score column corresponds with 97.72 percent. Your ice-cream cone is at the 97.72rd percentile, and 97.72 percent of the other single dip cones at Adrian's are smaller than yours. Lucky you!

11 Bob's commuting times to work have a normal distribution with a mean of 45 minutes and standard deviation of 10 minutes.

a. What percentage of the time does Bob get to work in 30 minutes or less?

b. Bob's workday starts at 9 a.m. If he leaves at 8 a.m., how often is he late?

12 Times to complete a statistics exam have a normal distribution with a mean of 40 minutes and standard deviation of 6 minutes. Deshawn's time comes in at the 90th percentile. What percentage of the students are still working on their exams when Deshawn leaves?

13 Suppose that your exam score has a standard score of 0.90. Does this mean that 90 percent of the other exam scores are lower than yours?

14 If a baby's weight is at the median, what's her percentile?

15 Clint sleeps an average of 8 hours per night with a standard deviation of 15 minutes. What's the chance he will sleep less than 7.5 hours tonight?

Suppose you know that Bob's test score is above the mean, but he doesn't remember by how much. At least how many students must score lower than Bob?

Finding Probabilities for a Normal Distribution

To find probabilities for any normal distribution:

1. **Convert the values to standard scores, using the Z-formula.**
2. **Look up their percentiles, using the Z-table (in the Appendix).**
3. **Use those percentiles (by adding them, subtracting them, taking them as they are, or taking 100 percent minus the percentile in the Z-table) to get your final answer.**

In this section, you practice each technique.

REMEMBER

You may come across many different types of tables for the Z-distribution; the Z-table I use (in the Appendix) is a common one, but the Z-table in your textbook may appear different and may even contain more decimal places than mine. My Z-table gives you the percentile, or area below a given value, not the area from the mean out to the value (as is done with the empirical rule [refer to Chapter 4] and some other types of Z-tables). So whether you should add or subtract probabilities after you find them depends on whether you're using the empirical rule (Chapter 4) or the Z-table (this is typically the case with most textbooks as well). Always make sure you understand exactly how your tables work before you attempt to solve probability problems.

See the following for an example of finding the probability of being between two values on a normal distribution.

EXAMPLE

Q. The weights of single dip ice-cream cones at Bob's ice cream parlor have a normal distribution with a mean of 8 ounces and standard deviation of one-half ounce (0.5 ounces). What's the chance that an ice-cream cone weighs between 7 and 9 ounces?

A. In this case, you want the area between two values, so you convert each of them to Z-scores, find their percentiles on the Z-table, and subtract those percentiles taking the largest one minus the smallest one. Nine ounces becomes $(9-8)\div 0.5 = 1\div 0.5 = 2$. The corresponding percentile for $Z = 2.0$ is 97.73. Seven ounces becomes $(7-8)\div 0.5 = -2$, which has a corresponding percentile of 2.28 from the Z-table. Subtracting these percentiles gives you the area between: $97.73 - 2.28 = 95.45\%$.

17 Bob's commuting times to work have a normal distribution with a mean of 45 minutes and standard deviation of 10 minutes. How often does Bob get to work in 30 to 45 minutes?

18 The times taken to complete a statistics exam have a normal distribution with a mean of 40 minutes and standard deviation of 6 minutes. What's the chance of Deshawn completing the exam in 30 to 35 minutes?

19 Times until service at a restaurant have a normal distribution with mean of 10 minutes and standard deviation of 3 minutes. What's the chance of it taking longer than 15 minutes to get service?

20 At the same restaurant as in Question 19 with the same normal distribution, what's the chance of it taking no more than 15 minutes to get service?

21 Clint, obviously not in college, sleeps an average of 8 hours per night with a standard deviation of 15 minutes. What's the chance of him sleeping between 7.5 and 8.5 hours on any given night?

22 One state's annual rainfall has a normal distribution with a mean of 100 inches and standard deviation of 25 inches. Suppose that corn grows best when the annual rainfall is between 100 and 150 inches. What's the chance of achieving this amount of rainfall?

Finding the Percentile (Backwards Normal)

Now you're ready to face the situation that students seem to dread the most and that professors seem sure to ask about. You're given the percentage of values that lie at the bottom or the top of the normal distribution, and you need to find the cutoff point that goes along with that percent. Want the good news? *Backwards normal* problems aren't as bad as they may seem, as long as you establish a pattern by which you can identify them when they come up and then work through them systematically. This section helps you do just that.

You know you have a backwards normal problem when you have a normal distribution, you are given a percentage of values above or below a cutoff point, and you need to find that cutoff point. To break out of backwards-normal purgatory:

1. **Identify a given percentile, using the given information.**
2. **Find the percentile's corresponding standard score, using the Z-table (in the Appendix).**
3. **Convert the Z-score back to original units.**

   To do this conversion, you can either take the Z-formula, put in the items you know (the mean, standard deviation, and value for *z*), and solve it for *x*, or you can use a formula that's already done that solving for you. I call this the Z-formula solved for *x*: $x = z\sigma + \mu$. It's the same as the Z-formula, except it has already been solved for *x* by cross multiplying by the standard deviation and adding the mean to both sides.

TIP

It's always a good idea to have the Z-formula and the Z-formula solved for x handy. That way you don't have to do all the algebra yourself if your instructor asks you to find x given a percentile. And you'll know right away which formula to use in which situation. If you're given x and asked for the percent, use the Z-formula. If you're given the percent and asked to find x, use the Z-formula solved for x. (*Note:* In the remaining problems, I use the Z-formula solved for x to solve backwards normal problems.)

To do a backwards normal problem, you literally work backward from how you find probabilities for the normal distribution. See the following for an example of a backwards normal problem, or reverse normal problem (as we call them in the business).

EXAMPLE

Q. Racehorses race around the track in a qualifying round according to a normal distribution with a mean of 120 seconds and standard deviation of 5 seconds. The top 10 percent of the times qualify; the rest don't. What's the cutoff time for qualifying?

A. You may think you need to find the 90th percentile, but no. The bigger values for times don't make the cut, and the smallest 10 percent of the times do make it, so in terms of data on a normal distribution, you know that the percentile of interest is the 10th. Remember that the percentage of times below the cutoff is 10, and the percentage above the cutoff is 90 (that's how percentiles work). The standard score for the 10th percentile is –1.28, looking at the Z-table. Converting this back to original units, you get $x = -1.28 * 5 + 120 = 120 - 6.4 = 113.6$. So the cutoff time is 113.6 seconds.

23 Weights have a normal distribution with a mean of 100 and standard deviation of 10. What weight has 60 percent of the values lying below it?

24 Jimmy walks a mile, and his previous times have a normal distribution with a mean of 8 minutes and standard deviation of 1 minute. What time does he have to make to get into his own top 10 percent of his fastest times?

25 The times it takes to complete a statistics exam have a normal distribution with a mean of 40 minutes and standard deviation of 6 minutes. Deshawn's time falls at the 42nd percentile. How long does Deshawn take to finish her exam?

26 Exam scores for a particular test have a normal distribution with a mean of 75 and standard deviation of 5. The instructor wants to give the top 20 percent of the scores an A. What's the cutoff for an A?

27 Service call times for one company have a normal distribution with a mean of 10 minutes and standard deviation of 3 minutes. Researchers study the longest 10 percent of the calls to make improvements. How long do the longest 10 percent last?

28 Statcars have a miles per gallon normal distribution with a mean of 75. Twenty percent of the vehicles get more than 100 miles per gallon. What's the standard deviation?

Answers to Problems in Normal Distribution

1. In the first normal distribution, most of the data falls between 10 and 70, each within 3 times 10 units of the mean (40), so you can assume that the standard deviation is around 10. You can also see that the saddle point (the place where the picture changes from an upside-down bowl to a right-side-up bowl) occurs about 10 units away from the mean of 40, which means the standard deviation is around 10. In the second normal distribution, the saddle point occurs about 5 units from the mean, so the standard deviation is at about 5. Notice also that most all the data lies within 15 units of the mean, which is 3 standard deviations.

2. See the following figure. Check that you have saddle points at 65 and 75.

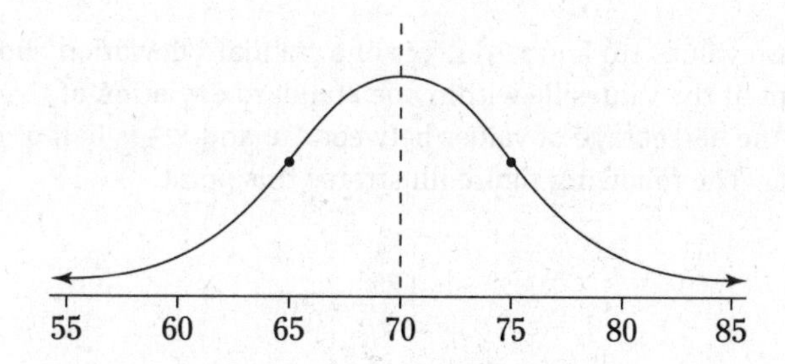

3. See the following figure. A graph with a standard deviation of 5 is taller and thinner than a graph whose standard deviation is 10. The normal distribution with a standard deviation of 10 is more spread out and flatter looking than the normal distribution with a standard deviation of 5, which looks more squeezed together close to the mean (which it is).

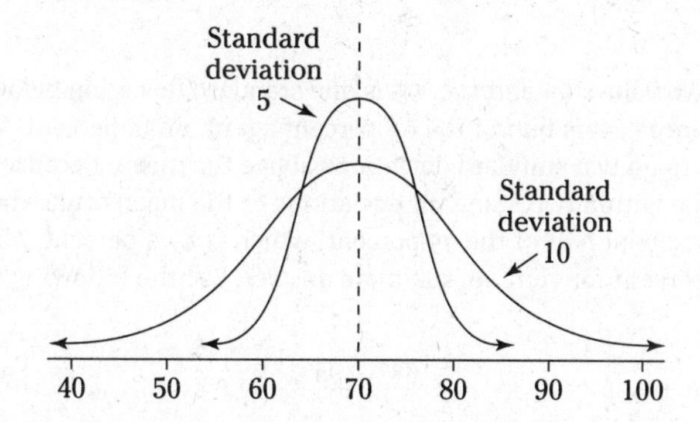

A normal distribution that looks flatter actually has more variability than one that goes from low to high to low as you look from left to right, because you measure variability by how far away the values are from the middle; more data close to the middle means low variability, and more data farther away means less variability. This characteristic differs from what you see on a graph that shows data over time (time series or line graphs). On the time graphs, a flat line means no change over time, and going from low to high to low means great variability.

4) A picture is worth a thousand points here (along with the empirical rule). See the following figure.

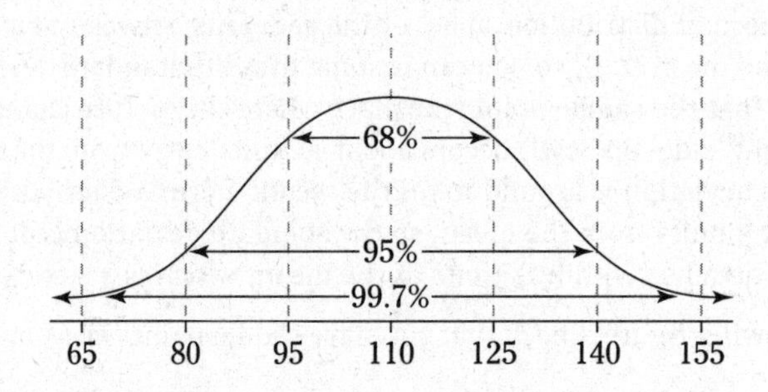

a. For the two values 110 and 125, 125 is one standard deviation above the mean, and about 68 percent of the values lie within one standard deviation of the mean (on both sides of it). So the percentage of values between 110 and 125 is half of 68 percent, which is 34 percent. The following figure illustrates this point.

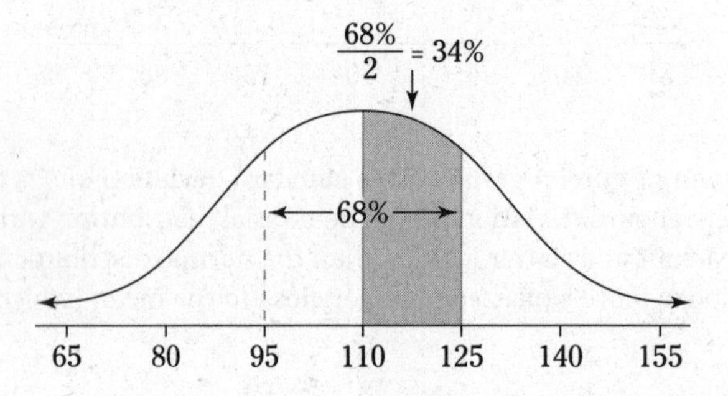

b. For the two values 95 and 140, 95 is one standard deviation below the mean (110), so that distance covers half of the 68 percent again, or 34 percent. To get from 110 to 140, you need to go two standard deviations above the mean. Because about 95 percent of the data lie within two standard deviations of the mean (on both sides of it), from 110 to 140 covers about half of the 95 percent, which is 47.5 percent. Add the 34 and the 47.5 to get 81.5 percent for your approximate answer. See the following figure for an illustration.

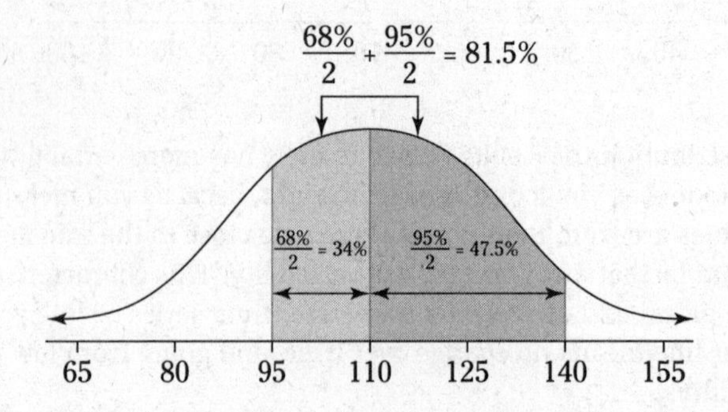

c. For the two values 80 and 95, 80 is two standard deviations below the mean of 110, which represents $95 \div 2$ or 47.5% of the data. And 95 is one standard deviation below 110, which represents $68 \div 2$ or 34% of the data. You want the area between these two values, so subtract the percentages: $47.5 - 34 = 13.5\%$. See the following figure for a visual.

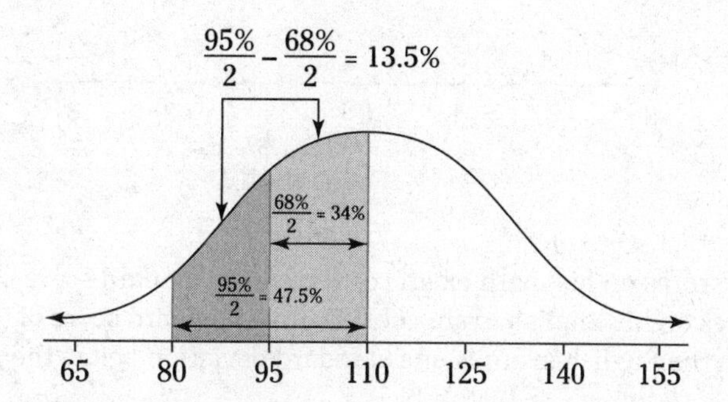

TIP

When using the empirical rule to find the percentage between two values, you may have to use a different approach, depending on whether both values are on the same side of the mean or one is above the mean and one below the mean. First, find the percentages that lie between the mean and each value separately. If the values fall on different sides of the mean, add the percentages together. When they appear on the same side of the mean, subtract their percentages (largest minus smallest, so the answer isn't negative). Or, better yet, draw a picture to see what you need to do.

5. The following figure shows a picture of this distribution. Take $80 - 70 = 10$ and divide by 10 to get 1. Bob's score is one standard deviation above the mean. You can see this on a picture as well, because Bob's score falls one "tick mark" above the mean on the picture of the original normal distribution.

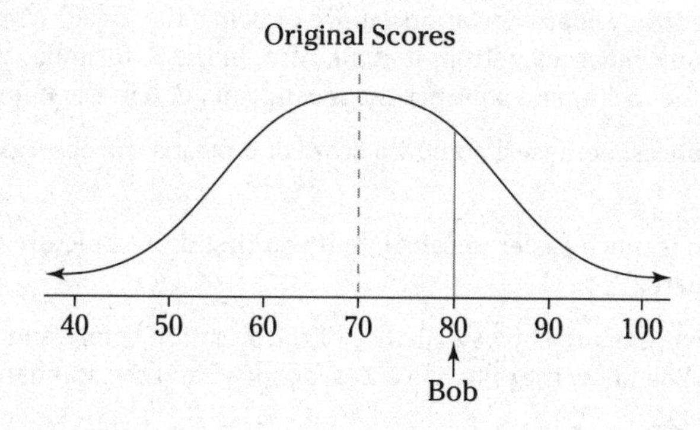

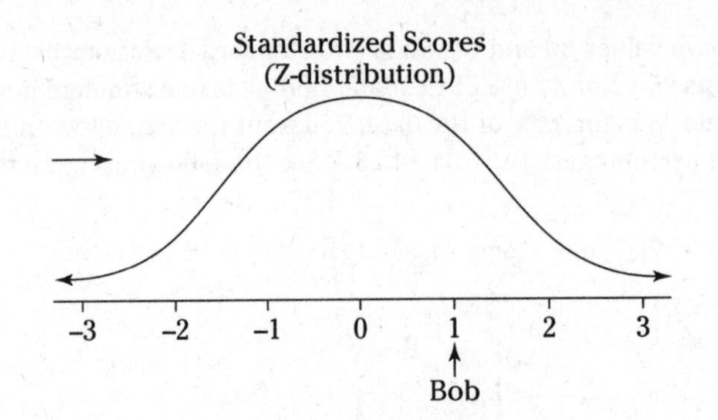

6. The 80 Bob scores on his math exam converts to a standard score of 1 (see Question 5). The 80 Bob scores on his English exam converts to a standard score of $(80-85)\div 5=-5\div 5=-1$. His score on the English exam is one standard deviation below the mean, so his math score is better.

Your actual scores don't matter; what matters is how you compare to the mean, in terms of number of standard deviations.

7. Here you can use the same formula, $z=\frac{x-\mu}{\sigma}$, but you need to plug in different values. You know that the mean is 70 and the standard deviation is 5. You know $Z=-2$, but you don't know x, the original score. So what you have looks like $-2=\frac{X-70}{5}$. Solving for X, you get $X-70=-2*5$, so $X-70=-10$, or $X=60$. The answer makes sense because each standard deviation is worth 5, and you start at 70 and go down two of these standard deviations: $70-(2*5)=60$. So Sue scored a 60 on the test.

8. A standard score of 0 means your original score is the mean itself, because the standard score is the number of standard deviations above or below the mean. When your score is on the mean, you don't move away from it at all. Also, in the Z-formula, after you take the value (which is at the mean) and subtract the mean, you get 0 in the numerator, so the answer is 0.

9. Exactly 20 ounces, because a standard score of 0 means the observation is right on the mean.

10. This problem is much easier to calculate if you first draw a picture of what you know and work from there.

    a. The following figure shows a picture of the situation before and after Fido's weight-loss program. You know that 150 has a Z-score of +2 and the standard deviation is 5, so you have $+2=\frac{150-\mu}{5}$. And solving for the mean (μ), you calculate $\mu=150-2(5)=140$. The mean weight for his breed/age group is 140 pounds.

    b. A Z-score of –2 corresponds to a weight 2 standard deviations below the mean of 150, which brings you down to $140-2(5)=130$. Fido weighs 130 pounds after the program.

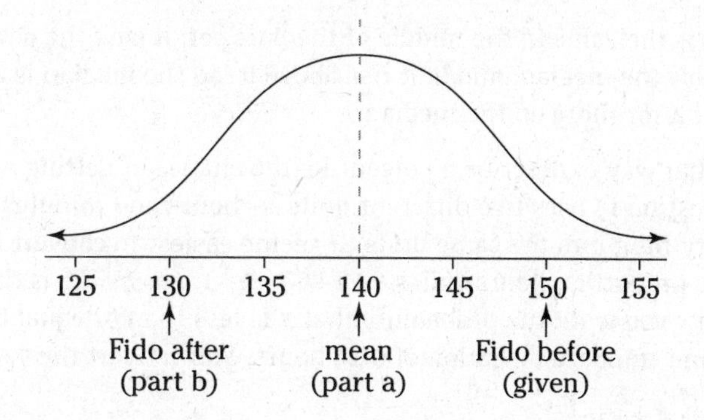

TIP

For any problem involving a normal distribution, drawing a picture is the key to success.

11. For this problem, you need to be able to translate the information into the right statistical task.

   a. Because you want a percentage of time that falls below a certain value (in this case, 30), you need to look for a percentile that corresponds with that value (30). You standardize 30 to get a Z-score of -1.5 by taking 30 minus 45 and then dividing by 10. This means a 30-minute commuting time is well below the mean (so it shouldn't happen very often). The percentile that corresponds to -1.5, according to the Z-table (see the Appendix), is 6.68, which means Bob gets to work in 30 minutes or less only 6.68 percent of the time.

   b. If Bob leaves at 8 a.m. and is still late for work, his commuting time must be more than 60 minutes (a time that brings him to work after 9 a.m.), so you want the percentage of time that his commute is above 60. Percentiles don't automatically give you the percentage of data that lies above a given number, but you can still find it. If you know what percentage of the time he gets to work in less than 60 minutes, you can take 100 percent minus that time to get the percentage of time he gets to work in more than 60 minutes. You standardize 60 to get $+1.5 = (60 - 45)$ divided by 10 and find the percentile that goes with that — 93.32%. Then take $100\% - 93.32\%$ to get 6.68%.

   The answer to part b of this question is the same as the answer to part a by symmetry of the normal distribution. Thirty and 30 are both 1.5 standard deviations away from the mean, so the percentage below 30 and the percentage above 60 should be the same.

12. Because Deshawn is at the 90th percentile, 90 percent of the students have exam times lower than hers, which means they all leave before the remaining 10 percent finish. So the answer is 10 percent.

13. No. It means your exam score is 0.90 standard deviations above the mean. You have to look up 0.90 on the Z-table (see the Appendix) to find its corresponding percentile, which is 81.59 percent; therefore, 81.59 percent of the exam scores are lower than yours.

REMEMBER

Make sure you keep your units straight; a Z-score between 0 and 1 looks a lot like a percentile, but it isn't!

(14) The median is the value in the middle of the data set; it cuts the data set in half. Half of the data fall below the median, and half rise above it. So the median is at the 50th percentile. (See Chapter 4 for more on the median.)

(15) Here's another way to describe a percentile: the chance of getting a value lower than a certain number. Question 15 uses two different units — hours and minutes — so the first step is to convert everything into the same units. It seems easiest to convert the 15 minutes to hours by using the proportion hours/minutes: $1 \div 60 = h \div 15$, so $h = .25$ is the standard deviation in minutes. Now you want the probability that X is less than or equal to 7.5, when x has a mean of 8 hours and standard deviation of 0.25 hours. You convert the 7.5 with the Z-formula to get $z = \frac{7.5 - 8}{.25} = \frac{-.5}{.25} = -2$.

Now look up the percentile because you want the probability of being less than that value. The answer is 2.28 percent (or 0.0228 in decimal form).

TIP

Setting up the problem correctly and knowing how to begin is 90 percent of the job. After you set it up, you shouldn't have a problem working it out. Spend time reading problems and thinking about how to start them for good practice.

(16) If Bob's score is above the mean, his standard score is positive, and the percentage of values below his score is above the 50th percentile. If his score is right at the mean, he scores in the 50th percentile.

(17) The following figure shows a picture of the situation. You want the probability that x is between 30 and 45, where x is the commuting time. Converting the 30 to a standard score with the Z-formula, you get $z = \frac{30 - 45}{10} = \frac{-15}{10} = -1.5$. Forty-five converts to zero because it's right at the mean. (Note that $45 - 45$ is 0, and 0 divided by 10 is still 0.) Because you have to find the probability of being between two numbers, you look up each of the percentiles associated with their standard scores on the Z-table (see the Appendix) and subtract their values (largest minus smallest, to avoid a negative answer). Using the Z-table, the percentile for $Z = -1.5$ is 6.68, and the percentile for $Z = 0$ is 50%. Subtracting those gives you $50 - 6.68 = 43.32\%$.

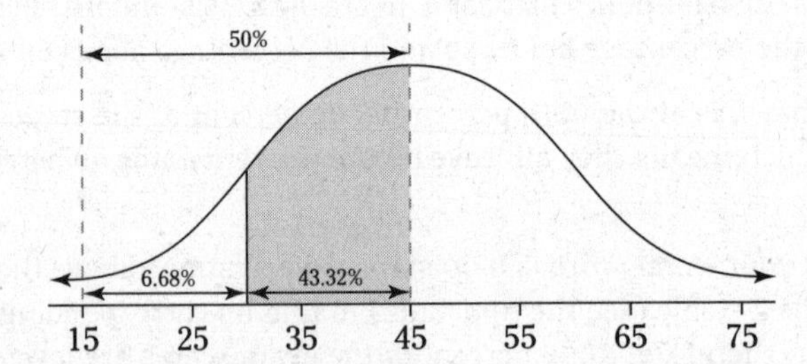

REMEMBER

The reason you subtract the two percentiles when finding the probability of being between two numbers is because the percentile includes all the probability less than or equal to a certain value. You want the probability of being less than or equal to the larger number, but you don't want the probability of being less than or equal to the smaller number. Subtracting the percentiles allows you to keep the part you want and throw away the part you don't want.

18 The following figure shows a picture of this situation. You want the probability that x (exam time) is between two values, 30 and 35, on the normal distribution. First, you convert each of the values to standard scores, using the Z-formula. The 30 converts to $z = \frac{30-40}{6} = \frac{-10}{6} = -1.67$ (or -1.7), which is at the 4.46th percentile. The 35 converts to $z = \frac{35-40}{6} = \frac{-5}{6} = -.83$ (or $-.8$), which is at the 21.19th percentile. To get the probability, or area, between the two values, subtract each of their percentiles — the larger one minus the smaller one — to get $21.19 - 4.46 = 16.73\%$.

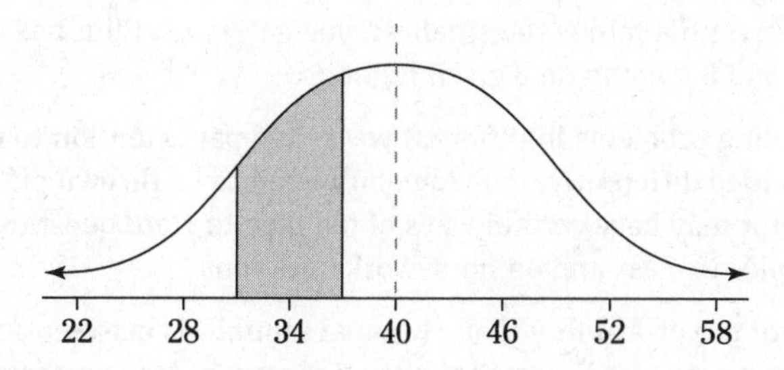

19 You want the probability that X (service time) is more than 15 minutes (see the following figure). In this case, you convert the value to a standard score, find its percentile, and take 100 minus the percentile because you want the percentage that falls above it. Substituting the values, you have $z = \frac{15-10}{3} = \frac{5}{3} = 1.67$ (or 1.7), which is at the 95.54th percentile. The probability you want is $100 - 95.54 = 4.46\%$.

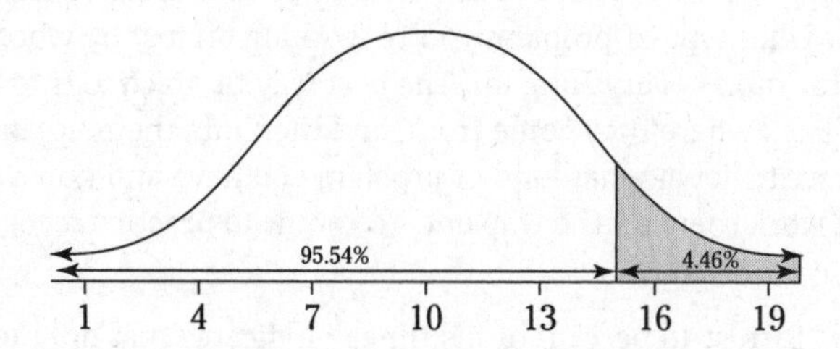

20 This problem is the exact opposite (or complement) of Question 19. Here, you want the probability that X is no more than 15, which means X is less than or equal to 15. If 4.46 percent (0.0446) is the probability of being more than 15 in Question 19, the probability of being less than or equal to 15 is $1 - 0.0446$, which is 0.9554 or 95.54%.

TIP

The chance of X being exactly equal to a certain value on a normal distribution is zero, so it doesn't matter whether you take the probability that *x* is less than 15 or the probability that *x* is less than or equal to 15. You get the same answer, because the probability that *x* equals 15 is zero.

21 For this problem, you need to convert 7.5 and 8.5 to standard scores, look up their percentiles on the Z-table (see Appendix), and subtract them by taking the largest one minus the smallest one. In this case, 7.5 converts to a standard score of -2 (see the answer to Question 15). Now 8.5 converts to a standard score of $+2$ because you have $z = \frac{8.5-8}{0.25} = \frac{+0.5}{0.25} = +2$. The corresponding percentiles for $Z = -2$ and $Z = +2$ are 2.28 and 97.72, respectively. Subtracting the largest percentile minus the smallest, you get 95.44. Clint has a high chance of sleeping between 7.5 and 8.5 hours on a given night.

TIP

Practice wording problems in different ways, and pay attention to homework and in-class examples worded differently, too. You don't want to be thrown off in an exam situation. Your instructor may have certain ways of her own to word questions. This comes through in her examples in class and on homework questions.

22 Here you want the probability that *x* (annual rainfall) is between 150 and 100, so convert both numbers to standard scores with the Z-formula, look up their percentiles, and subtract (take the biggest minus the smallest). The 150 converts to $+2$ and has a percentile of 97.72. The 100 converts to 0 and has a percentile of 50. Subtract the percentiles to get the area between them: $97.72 - 50 = 47.72\%$.

23 This problem essentially asks for the score that corresponds to the 60th percentile (the tricky part is recognizing this). First, you look up the standard score for the 60th percentile, which is 0.25. Using the Z-formula, you solve for *x* and get $x = .25(10) + 100 = 102.5$.

TIP

When you study material, your books all organize the topics by chapter and section, so you already know what type of problem you're working on just by where it is. But on an exam, your instructor mixes everything up. The best way to practice is to make copies of problems, write on the back where they come from, and then mix them up and put them in a pile. Go through and write down what kind of problem you have and how you should start each one — don't work them all the way out. You want to practice recognizing the problem and starting it correctly.

24 The wording "fastest 10 percent of his times" indicates that only 10 percent of his times are *less* than his current one; therefore, the percentile is 10 (not 90), and the standard score is -1.28 from the Z-table in the Appendix (you find the percentile closest to 10 and look at the standard score). Converting back to original units (*x*) and using the Z-formula to solve for *x*, you have $-1.28(1) + 8 = 6.72$ minutes. Jimmy has to walk in 6.7 minutes to get into his top 10 percent.

25 You know the percentile, and you want the original score. The middle step is to take the percentile, find the standard score that goes with it, and then convert to the original score (*x*) with the Z-formula solved for *x*. In this case, you have a percentile of 42, and the table in the Appendix tells you the standard score is about -0.2. Using the Z-formula solved for *x*, you have $x = (-0.2)(6) + 40$, which gives you 38.8. It takes Deshawn about 38.8 minutes to finish the exam.

26 Because the top 20 percent of scores get As, the percentage below the cutoff for an A is $100-20=80\%$. Coincidentally, the standard score corresponding to the 80th percentile is about 0.84. Converting 0.84 to original units (x) with the Z-formula solved for x, you have $x=(0.84)(5)+75=79.2$. The cutoff for an A is 79.

REMEMBER

Whatever Z-table you use, make sure you understand how to use it. The Z-table in this workbook doesn't list every possible percentile; you need to choose the one that's closest to the one you need. And just like an airplane where the closest exit may be behind you, the closest percentile may be the one that's lower than the one you need.

27 The longest 10 percent are the calls with the 10 percent biggest values. The lengths of calls below the cutoff is 90 percent, which is the 90th percentile (percentile is the area below the value). The corresponding Z-score is 1.28, which converts to $x=(1.28)(3)+10=13.84$ with the Z-formula solved for x. So the cutoff for the 10 percent longest customer service calls is 13.84 minutes.

TIP

You may have thought that the Z-score here would be -1.28 because it corresponds to the 10th percentile. But remember, the longer the phone call, the larger the number will be, and you're looking for the longest phone calls, which means the top 10 percent of the values. Be careful in testing situations; these kinds of problems are used very often to make sure you can decide whether you need the upper tail of the distribution or the lower tail of the distribution.

28 I saved the best for last! I consider this problem a bit of an advanced, extra-credit type exercise. It gives you a percent, but it asks you for the standard deviation. When in doubt, work it like all the other problems and see what materializes. You know the percentage of these vehicles getting more than 100 miles per gallon is 20, so the percentage getting less than 100 is 80 (the percentile). The standard score corresponding to the 80th percentile is 0.84 (see the Appendix). You know the standard score, the mean, and the value for x in original units (100). Put all this into the Z-formula solved for x to get $100=0.8\sigma+75$. Solving this formula for σ (the standard deviation) gives you $\sigma=\frac{100-75}{0.8}=\frac{25}{0.8}=31.25$. The standard deviation is 31.25 mpg.

[illegible]

[illegible]

[illegible]

[illegible]

[illegible]

[illegible]

[illegible]

» Finding probabilities using the formula and table

» Determining the mean and variance

» Estimating probabilities in large cases with the normal approximation

Chapter 7
The Binomial Distribution

In this chapter, you find out how the binomial distribution helps you work with data that has only two outcomes: yes or no — for example, whether someone owns a cellphone or whether a certain drug worked for a disease. You see how to distinguish the binomial from other types of distributions, and you find probabilities for a binomial in small, medium, and large sample cases.

Characterizing the Binomial Distribution

The term *binomial* means "two names," which comes from the fact that each individual observation in this situation has only one of two possible outcomes: yes or no. The binomial random variable counts the number of yeses that occur among all the observations. For example, if you flip a coin ten times, each flip results in either a head (which you may call a yes) or a tail (which you may call a no). The binomial random variable counts the number of yeses, or heads, that occur among the ten flips, meaning the variable itself can take on the values 0 (no heads) all the way to 10 (all heads). Each of these counts has its own probability, and when you put all the probabilities together, you have a binomial distribution with n observations (also known as trials) and p as the probability of a yes. (In the case of flipping a coin ten times, n is 10, and p is 0.50.) The binomial distribution gives each possible count of the number of yesses (out of n observations), along with the probability associated with each count.

To get more specific, every binomial distribution has four certain characteristics that all must occur; they are as follows:

- A fixed number (n) of observations or trials occur.
- Each observation is independent of all the others.
- Each observation has two possible outcomes: yes or no (also known as success or failure).
- The probability of yes is p and remains the same for each observation.

See the following for an example of assessing whether you have a binomial distribution.

EXAMPLE

Q. Suppose that you randomly select M&Ms from a large bag until you have five red ones. You count the number of tries that it takes for this to occur. Is this a binomial distribution?

A. No. You do not have a fixed number of observations, n. You do not know how many trials it will take until five red M&Ms are drawn.

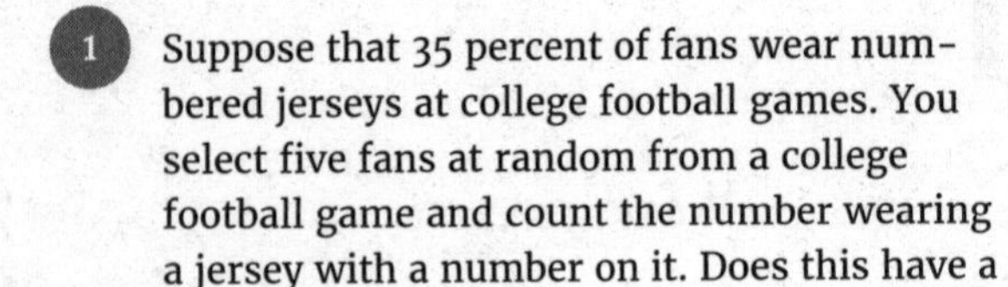

1. Suppose that 35 percent of fans wear numbered jerseys at college football games. You select five fans at random from a college football game and count the number wearing a jersey with a number on it. Does this have a binomial distribution?

2. Suppose that you randomly guess your answers to all the questions on a 20-question true/false test. Does the number of correct answers have a binomial distribution?

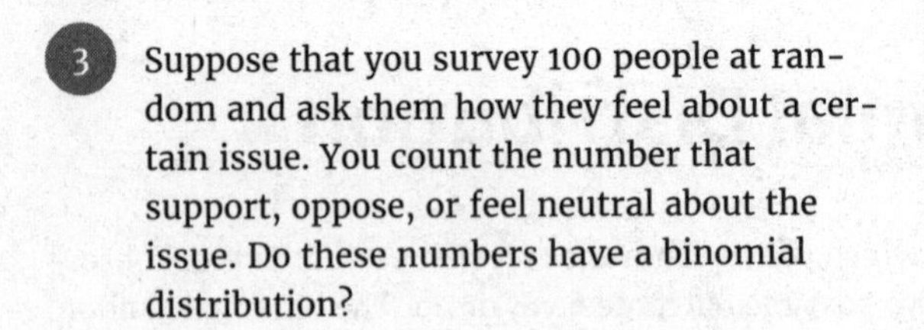

3. Suppose that you survey 100 people at random and ask them how they feel about a certain issue. You count the number that support, oppose, or feel neutral about the issue. Do these numbers have a binomial distribution?

4. You have a group of five women and five men, and you need to choose three to be on a committee. You choose three people at random. Does the number of women chosen for the committee have a binomial distribution?

Finding Probabilities Using the Binomial Formula for small n

With a binomial distribution, you let X represent the number of yeses among the n observations. X can take on the values of 0, 1, 2, . . . n. To find the probability of X yeses among the n observations, you use the following formula:

$$P(x) = \binom{n}{x} p^x (1-p)^{n-x}, \text{ where } \binom{n}{x} = \frac{n!}{x!(n-x)!} \text{ and } n! = n(n-1)(n-2)\ldots(3)(2)(1).$$

For example, $5! = (5)(4)(3)(2)(1) = 120$; $2! = (2)(1) = 2$; $1! = 1$; and 0! by definition is 1.

Items to note about the formula:

- p is the probability of yes (success).
- $(1-p)$ is the probability of no (failure).
- x is the number of yeses among the n observations.
- $(n-x)$ is the number of noes among the n observations.

EXAMPLE

Q. If you flip a coin five times, what's the chance you will get three heads?

A. In this case, x = number of heads has a binomial distribution with $p = 0.5$ (chance of a head) and $n = 5$ flips. You want the probability that $x = 3$, which is P(3). Using the formula, you find

$$P(3) = \binom{5}{3} 0.5^3 (1-0.5)^{5-3} = \frac{5!}{3!(5-3)!} 0.5^5$$

$$= \frac{5*4*3*2*1}{(3*2*1)(2*1)}(0.03125) = 0.3125.$$

5 A stoplight is green 40 percent of the time. If you stop at this light eight random times, what is the chance that it's green exactly five times?

6 If 10 percent of the parts made by a certain company are defective and have to be remade, what is the chance that a random sample of four parts has one that is defective?

7 If 40 percent of university students purchase their textbooks online, in a random sample of five students, what's the chance that exactly one of them purchased their textbooks online?

8 Suppose that 60 percent of families own a pet. You randomly sample four families. What is the chance that two or three of them own a pet?

9 Your chance of winning a small prize in a scratch-off ticket is 10 percent. You buy five tickets. What's the chance you will win at least one prize?

10 Suppose that 80% of athletes at a certain college graduate. You randomly select eight athletes. What's the chance that at most 7 of them graduate?

Finding Probabilities Using the Binomial Table for Medium-Sized n

Some probabilities for the binomial have already been calculated and are available in table form for selected values of p and n. The table comes in handy if you need multiple probabilities to solve a problem or if the n value is larger than you might want to work with using the formula. A binomial table for this book is located in the Appendix.

Follow these steps to using the binomial table to find probabilities for a binomial distribution with n observations (trials) and p = probability of yes (success):

1. **Find the minitable associated with your particular value of *n*.**
2. **Locate the column that represents your particular value of *p* (or the one closest to it, if appropriate).**
3. **Find the row that represents the number of yeses (successes), *x*, that you are interested in.**
4. **Intersect the row and column from Steps 2 and 3. This gives you the probability $p(x)$ for *x* successes.**

See the following for how to find a binomial probability using the binomial table.

EXAMPLE

Q. Suppose that x is binomial with $n = 10$ and $p = 0.9$. What's the probability that $X = 8$?

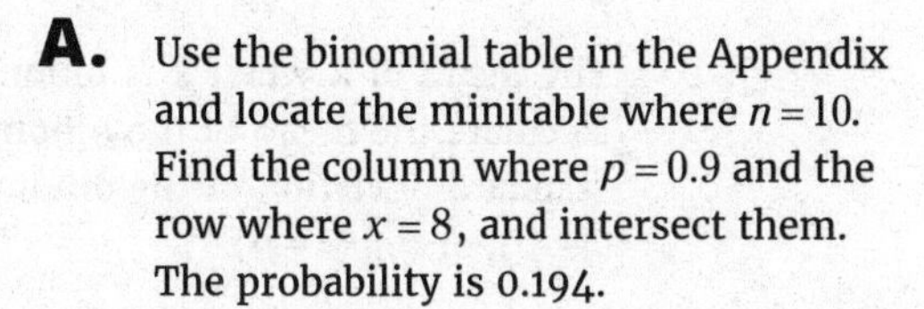

A. Use the binomial table in the Appendix and locate the minitable where $n = 10$. Find the column where $p = 0.9$ and the row where $x = 8$, and intersect them. The probability is 0.194.

11 Bob's commuting times to work are varied. He makes it to work on time 80 percent of the time. On 12 randomly selected trips to work, what's the chance that Bob makes it on time at least 10 times?

12 Suppose that you flip a fair coin four times. What's the chance of getting at least one head?

13 Suppose that the chance that an elementary student eats hot lunch is 30 percent. What's the chance that, among 20 randomly selected students, between 6 and 8 students eat hot lunch (inclusive)?

Calculating the Mean and Variance of the Binomial

The mean of X when X is binomial is the number of yeses you expect to get in n trials. You calculate the mean of the binomial by finding n times p, or np. The variance is $np(1-p)$. The standard deviation of the binomial is calculated by taking the square root of n times p times $(1-p)$, or $\sqrt{np(1-p)}$.

In this section, you practice finding the mean and standard deviation of the binomial. See the following for an example.

EXAMPLE

Q. Suppose that X has a binomial distribution with $n = 10$ and $p = 0.6$. Find the mean, variance, and standard deviation of X.

A. The mean in this case is $np = 10 * 0.6 = 6$; the variance is $10 * 0.6 * 0.4 = 2.4$; and the standard deviation is the square root of 2.4, which is 1.55.

14 Suppose that you survey a random sample of 100 people and count the number who own at least one pet (call this cont X). If it is known that 50 percent of people own at least one pet, find the mean, variance, and standard deviation of X.

15 Suppose that a multiple-choice test has ten questions and four choices for each question. Let X represent the number of correct answers if the person was randomly guessing on each question. Find the mean, variance, and standard deviation of X.

REMEMBER

The mean and standard deviation of x are in the same units as x. The variance uses squares, so you typically do not include units with it.

Estimating Probabilities in Large Cases — the Normal Approximation

In cases where n is large enough (for example, too large to use the binomial table), you can try to use the normal distribution to approximate your answer (for more info on the normal distribution, see Chapter 6).

1. **Check two conditions to be sure n is large enough: $np \geq 10$ and $n(1-p) \geq 10$. If both conditions are met, proceed to Step 2. If one or both are not met, you can't use the normal approximation.**
2. **Find the mean of X by calculating np, and call it μ.**
3. **Find the standard deviation of X by calculating $\sqrt{np(1-p)}$, and call it σ.**
4. **Convert X to a standard score, using the Z-formula, $Z = \frac{X-\mu}{\sigma}$ (see Chapter 6).**
5. **Use the Z-table to find the final answer (see the Appendix).**

In this section, you practice the normal approximation to solve binomial problems when n is large.

EXAMPLE

Q. It is estimated that 70 percent of people have some type of social media account. You take a random sample of 100 people. What's the chance that at least 75 of them have some type of social media account?

A. First, note that $n = 100$, and $p = 0.70$, and you want $P(X \geq 75)$. Checking the two conditions in Step 1, you have $np = 100(0.70) = 70$ is at least 10, and $n(1-p) = 100(1-0.70) = 30$ is at least 10, so you're good to go with the normal approximation. Next, in Step 2, you find $np = 100(0.70) = 70 = \mu$. Then in Step 3, you calculate $np(1-p) = 100(0.70)(0.30) = 21$ and take the square root, which is $4.58 = \sigma$. Now in Step 4, you convert X to $Z = \frac{X-\mu}{\sigma} = \frac{75-70}{4.58} = 1.09$. You want $P(X \geq 75) = P(Z \geq 1.09) = 1 - 0.8621 = 0.1379$ from the Z-table in the Appendix. This is an approximate answer.

TIP

If you had not used the normal approximation, you would have had to calculate $P(X \geq 75) = P(X = 75) + P(X = 76) + P(X = 77) + \ldots + P(X = 100)$, which would take a lot of work! If you find yourself doing a long string of probabilities by hand, consider the normal approximation, but be sure to check the conditions first.

16 Suppose that 90 percent of student athletes at a certain university graduate. You take a random sample of 200 athletes. What is the chance that at most 180 of them will graduate?

17 Suppose that 20 percent of grocery store customers use coupons. What is the chance that, out of 50 randomly selected grocery store customers, more than 15 of them use coupons?

18 Suppose that 80 percent of college students report to have worked at a restaurant. You survey 100 randomly selected college students. What's the chance that fewer than 75 have worked at a restaurant?

19 Suppose that Bob is a U.S. senator, and his approval rating is 70 percent. If you randomly survey 100 of Bob's constituents, what's the chance that more than 75 of them will approve of Bob?

Answers to Problems in the Binomial Distribution

1. Yes. It has all four elements of a binomial distribution. First, it has n observations (5); the value $p = 0.35$ as the chance of a yes (success) given in the problem as the chance of a yes (success). The results are independent because the people are chosen at random, and each observation has a yes or no outcome because the fan either is or is not wearing a numbered jersey.

2. A true/false test has two outcomes on each question: correct or incorrect. The chance of getting the problem right by guessing is 50 percent. You're counting the number of correct answers out of 20 questions. Because a true/false test has two outcomes, the chance of a correct answer (success) is $1/2 = 0.50$. Randomly guessing means that the results for each question are independent. So if X counts the number of correct answers out of 20, then X is binomial.

REMEMBER

Sometimes p is not given in the problem, and you have to figure it out. In this case, the questions have two options — true or false — so the chance of being right is 50 percent.

3. This is not a binomial situation, because there are three possible outcomes for each observation (person surveyed): support, oppose, or neutral. For a binomial, there can be only two possible outcomes on each observation.

4. No, this is not a binomial. Before you choose the first person for the committee, the chance of choosing a woman is 5/10 or 0.50. But say that a woman was chosen, then the probability changes to 4/9, and if you don't choose a woman, the probability changes to 5/9. The main point is, there are only 9 people left once you've chosen one person, so the value of p (here, the chance of choosing a woman) changes. You would have to put the people back once they were chosen to not change the probability, but that's not going to help you choose your committee.

5. This is a small case of a binomial situation with $n = 8$ observations and $p = 0.40$ as the chance of success (light is green.) You want $P(X = 5)$ where X is the number of times the light is green out of the eight observations. You can use the binomial formula for this one:

$$P(X=5)=\binom{8}{5}0.40^5(1-0.40)^{8-5}=\frac{8!}{5!(8-5)!}0.4^5 0.6^3=\frac{8*7*6*5!}{5!(3*2*1)}(0.0022)=56*0.0022=0.12$$

You could use the binomial table to do this problem (see next answer), but be ready for your instructor to insist that you use the binomial formula to do some of the problems; problems like these with relatively small values of n are good practice.

6. This is also a binomial situation, where $n = 4$ and $p = 0.10$. You want $P(X = 1)$, where X is the number of defective parts found in the sample. You can use the binomial formula and get

$$P(X=1)=\binom{4}{1}0.1^1(1-0.1)^{4-1}=\frac{4!}{1!(4-1)!}0.1*0.9^3=\frac{4*3*2*1}{1(3*2*1)}0.0729=0.29.$$

(7) This is a binomial problem with $n = 5$ and $p = 0.40$. You want $P(X = 1)$, so you use the binomial formula:

$$P(X=1)=\binom{5}{1}0.4^1(1-0.4)^{5-1}=\frac{5!}{1!(5-1)!}0.4^1*0.6^4=\frac{5*4*3*2*1}{(1)(4*3*2*1)}0.0518=0.26$$

(8) This is a small case binomial problem with $n = 4$ and $p = 0.60$. You want $P(X = 2 \text{ or } 3) = P(X = 2) + P(X = 3)$. Using the binomial formula twice and summing the results, you get the following:

$$P(X=2)=\binom{4}{2}.6^2(1-.6)^{4-2}=\frac{4!}{2!(4-2)!}.6^2*.4^2=\frac{4*3*2*1}{(2*1)(2*1)}.0576=0.35$$

$$P(X=3)=\binom{4}{3}0.6^3(1-0.6)^{4-3}=\frac{4!}{3!(4-3)!}0.6^3*0.4^1=\frac{4*3*2*1}{(3*2*1)(1)}0.0864=0.35$$

$$P(X=2)+P(X=3)=0.35+0.35=0.70$$

(9) In this case, you have a binomial with $n = 5$ and $p = 0.10$. The probability of at least one prize is the same as the probability of one or more prizes (in other words, 1, 2, 3, 4, or 5). This is the same as 1 minus the chance of getting 0 tickets, which is easier to find. Using the binomial formula, you find

$$P(X\geq 1)=1-P(X=0)=1-\binom{5}{0}0.1^0(1-0.1)^{5-0}=1-\frac{5!}{0!(5-0)!}1(0.9^5)=1-0.59=0.41.$$

(10) Here, you have a binomial with $n = 8$ and $p = 0.80$. You want the probability that X is at most 7. This is the same as saying the highest value is 7 or that the possible values you need are 0 through 7. Again, the complement rule comes to the rescue and makes the problem easier because you can take 1 minus the probability that $X = 8$ and be done with it. Using the binomial formula, you get

$$P(X\leq 7)=1-P(X=8)=1-\binom{8}{8}0.8^8(1-0.8)^{8-8}=1-\frac{8!}{8!(8-8)!}0.1678(1)=1-0.17=0.83.$$

(11) This binomial has $n = 12$ and $p = 0.80$. You want $P(X \geq 10) = P(X = 10) + P(X = 11) + P(X = 12)$. Find each of these values in the binomial table where the minitable is $n = 12$, the column is 0.80, and the rows (marked by values of X) are 10, 11, and 12. These three numbers are $0.283 + 0.206 + 0.069 = 0.558$.

(12) When you flip a coin and count the number of heads (successes), you have a binomial with n = number of flips (in this case, 4) and $p = 0.50$, assuming that the coin is fair because the probability of a head is 1/2. You want $P(X \geq 1) = 1 - P(X = 0)$, so in the binomial table, find the minitable for $n = 4$ and the row for $p = 0.50$. Then look at the row for $X = 0$ (this will be row 1), and you find 0.063. This is $P(X = 0)$. So the final answer is $1 - P(X = 0) = 1 - 0.063 = 0.937$.

(13) In this problem, you have a binomial with $n = 20$ and $p = 0.30$. You want $P(6 \leq X \leq 8) = P(X = 6) + P(X = 7) + P(X = 8)$. Go to the binomial table and the minitable where $n = 20$ and the column where $p = 0.30$. Find the values for the rows where $X = 6$, 7, and 8, and sum them up to get $0.192 + 0.164 + 0.114 = 0.47$.

(14) This is a binomial with $n = 100$ and $p = 0.50$. The mean is $np = 100 * 0.50 = 50$ (people); the variance is $np(1 - p) = 100 * 0.50 * 0.50 = 25$; the standard deviation is the square root of 25, which is 5 (people).

15. Because there are ten questions with four choices each, you have a binomial with $n = 10$ and $p = 0.25$ as the chance of being correct (success); the mean is $np = 10 * 0.25 = 2.5$ questions; the variance is $np(1-p) = 10 * 0.25 * 0.75 = 1.875$; and the standard deviation is the square root of 1.875, which is 1.37 (questions).

16. In this case, $n = 200$ and $p = 0.90$, and it's a binomial scenario. Because n is large, you check the conditions of the normal approximation: $np = 200 * 0.90 = 180$ is at least 10, and $n(1-p) = 200 * 0.10 = 20$ is at least 10. Next in Step 2, you find $np = 200(0.90) = 180 = \mu$. Then in Step 3, you calculate $np(1-p) = 200(0.90)(0.10) = 18$ and take the square root, which is $4.24 = \sigma$. Now in Step 4, you convert X to $Z = \frac{X-\mu}{\sigma} = \frac{180-180}{4.24} = 0$. You want $P(X \leq 180) = P(Z \leq 0) = 0.50$ using the Z-table in the appendix. This is an approximate answer.

TIP

This is an approximate answer because the problem is really a binomial distribution problem, and you're using a normal distribution to solve it. As long as the conditions are met, that means n is large enough for the normal approximation to work okay. And the larger n is, the better the approximation works.

17. Here, you have a binomial distribution with $n = 50$ and $p = 0.20$, and you want $P(X > 15)$. Because n is large, you check the conditions of the normal approximation: $np = 50 * 0.2 = 10$ is at least 10, and $n(1-p) = 50 * 0.8 = 40$ is at least 10. Next in Step 2, you find $np = 50 * 0.2 = 10 = \mu$. Then in Step 3, you calculate $np(1-p) = 50(0.20)(0.80) = 8$ and take the square root, which is $2.83 = \sigma$. Now in Step 4, you convert X to $Z = \frac{X-\mu}{\sigma} = \frac{15-10}{2.83} = 1.77$. You want $P(X > 15) = P(Z > 1.77) = 1 - 0.9616 = 0.0384$ using the Z-table in the Appendix. This is an approximate answer.

18. This is a binomial distribution with $n = 100$ and $p = 0.80$, and you want $P(X < 75)$. Because n is large, you check the conditions of the normal approximation: $np = 100 * 0.8 = 80$ is at least 10, and $n(1-p) = 100 * 0.20 = 20$ is at least 10. Next in Step 2, you find $np = 100 * 0.8 = 80 = \mu$. Then in Step 3, you calculate $np(1-p) = 100(0.8)(0.2) = 16$ and take the square root, which is $4 = \sigma$. Now in Step 4, you convert X to $Z = \frac{X-\mu}{\sigma} = \frac{75-80}{4} = -1.25$. You want $P(X < 75) = P(Z < -1.25) = 0.1056$ using the Z-table in the appendix. This is an approximate answer.

19. The number of constituents who approve of Bob in the sample has a binomial distribution with $n = 100$ and $p = 0.70$. You want $P(X > 75)$. Because n is large, you check the conditions of the normal approximation: $np = 100 * 0.7 = 70$ is at least 10, and $n(1-p) = 100 * 0.30 = 30$ is at least 10. Next in Step 2, you find $np = 100 * 0.7 = 70 = \mu$. Then in Step 3, you calculate $np(1-p) = 100(0.7)(0.3) = 21$ and take the square root, which is $4.58 = \sigma$. Now in Step 4, you convert X to $Z = \frac{X-\mu}{\sigma} = \frac{75-70}{4.58} = 1.09$. You want $P(X > 75) = P(Z > 1.09) = 1 - 0.8621 = 0.14$ using the Z-table in the Appendix. This is an approximate answer.

» Finding probabilities using the *t*-table

Chapter 8
The *t*-Distribution

The *t*-distribution is a very commonly used distribution in data analysis. If you've heard of the *t*-test (and if you haven't, you will), it's the *t*-distribution that plays a big part. In this chapter, you find out how the *t*-distribution works, how to find probabilities for it, and how it relates to the Z-distribution (see Chapter 6).

Getting to Know the *t*-Distribution

The *t*-distribution is typically used to study the mean of a population. It is often used to estimate the population mean: for example, the average age of the people who buy a certain product, or the average price of gas in your hometown. It is also often used to test the reported value of a population mean: For example, someone may want to know whether the average age of people buying the product is higher this year than last year, or someone may report that the average price of gas is a certain value and you want to challenge, or test, this reported value.

The *Z*-distribution (discussed in Chapter 6) is also used to answer these kinds of questions, and in fact, the *Z*- and *t*-distributions can be thought of as cousins. The idea here is to sort out which particular role the *t*-distribution plays and how it relates to the *Z*-distribution.

To get specific, every *t*-distribution has certain characteristics:

- It is a continuous distribution, going from negative infinity to positive infinity.
- It is symmetric, with a mound in the middle and tails sloping downward on each side.

» Its mean is in the middle and is equal to 0.

» Its standard deviation is denoted by σ.

These characteristics are all similar to some normal distributions (see Chapter 6). However, one characteristic separates the t-distribution from the normal and, in particular, the Z-distribution. The speed by which the sides slope down on the t-distribution depends on the sample size (whereas, with the normal, it's always the same — most of the values lie within three standard deviations of the mean.) Their shapes are flatter; their values are more spread out. That's because results based on less data are more variable than results based on more data.

Each t-distribution is distinguished by what statisticians call its *degrees of freedom*. In situations where you have one population being studied and the sample size is n, the degrees of freedom for the t-distribution is $n-1$. The notation for a t-distribution is the letter t with the degrees of freedom in the subscript. For example, the t-distribution with 3 degrees of freedom is denoted t_3.

REMEMBER

The larger n is, the more the t-distribution looks like a Z-distribution. At the point where the degrees of freedom is about 30, the t-distribution and the Z-distribution look about the same.

Figure 8-1 shows what different t-distributions look like for different sample sizes and how they all compare to the standard normal (Z-) distribution.

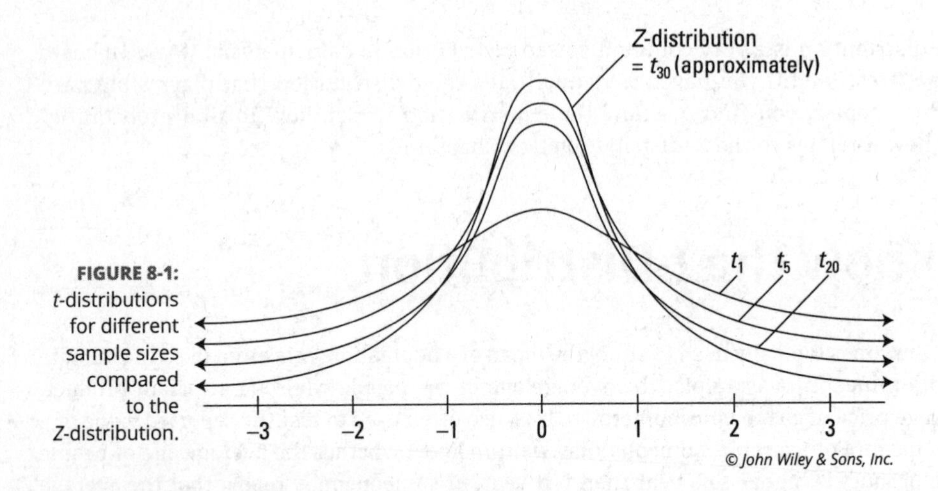

FIGURE 8-1: t-distributions for different sample sizes compared to the Z-distribution.

See the following for an example of characterizing the t-distribution.

EXAMPLE

Q. A t-distribution based on samples of size ten has how many degrees of freedom?

A. This t-distribution has $n-1=10-1=9$ degrees of freedom.

1 A t-distribution based on samples of size 25 has how many degrees of freedom?

2 What sample size is affiliated with the t_{17} distribution?

3 Which t-distribution is flatter, t_3 or t_{10}?

4 Which t-distribution looks more like the Z-distribution, t_{15} or t_{30}?

Working with the *t*-Table and Degrees of Freedom

Some probabilities have values already calculated on each *t*-distribution with degrees of freedom from 1 to 30 and are available in table form in what is called the *t*-table (see the Appendix). The *t*-table shows what statisticians call right-tail probabilities, which means the probability greater than a certain value in the right side of the *t*-distribution.

Follow these steps to using the *t*-table to find right-tail probabilities for *t*-distribution with $n-1$ degrees of freedom:

1. **Find the *t*-value for which you want the right-tail probability (call it *t*), and find the sample size that you're dealing with.**

   When studying one population mean, the sample size is *n*.

2. **Find the row corresponding to the degrees of freedom (*df*) for your problem (for example, when studying one population mean, $df = n - 1$). Follow that row across to find the two *t*-values between which your value of *t* falls.**

3. **Go to the top of the columns containing the two *t*-values from Step 2.**

   The right-tail (greater-than) probability for your *t*-value is somewhere between the two values at the top of these columns.

REMEMBER

The second-to-last row of the *t*-table represents the right-tail probabilities for the *Z*-distribution, because the *t*-distribution approaches a *Z*-distribution the larger *n* gets.

See the following for how to find a right-tail (greater-than) probability using the *t*-table.

EXAMPLE

Q. Suppose that *t* is 3.95 and *n* is 6. What is the corresponding right-tail probability on the *t*-table?

A. Using the *t*-table (see the Appendix), follow the steps. First, the value of *t* is 3.75 and *n* is 6, so you find the row corresponding to degrees of freedom $(df) = n - 1 = 6 - 1 = 5$. Then follow the row across and see that $t = 3.95$ lies between 3.36 and 4.03. Next, go to the top of their corresponding columns, and you find the right-tail probability for 3.95 lies between 0.01 and 0.005. That means the probability of observing a value of 3.75 or greater is between 0.005 and 0.01.

5 Suppose that t is 2.00 and n is 24. What is the corresponding right-tail probability on the t-table?

6 What is the right tail probability for $t = 1.00$ where n is 5?

7 What is the value of t such that the degrees of freedom are 10 and the right-tail probability is 0.05?

8 What is the right-tail probability for $t = 1.96$ with degrees of freedom 30?

9 If you're studying the population mean with a sample of size 100 and your t-value is 2.00, what would be the approximate right tail-probability?

10 What is the probability that t_9 is less than 0.702722?

Answers to Problems in the *t*-Distribution

1) A t-distribution based on samples of size 25 has $n-1=25-1=24$ degrees of freedom.

2) The sample sizes affiliated with t_{17} distribution is 18 because $df=n-1$ means $17=n-1$ and n must be 18.

3) t_3 is flatter than t_{10} because it has fewer degrees of freedom.

The lower the degrees of freedom, the flatter the t-distribution is and the more variability it has.

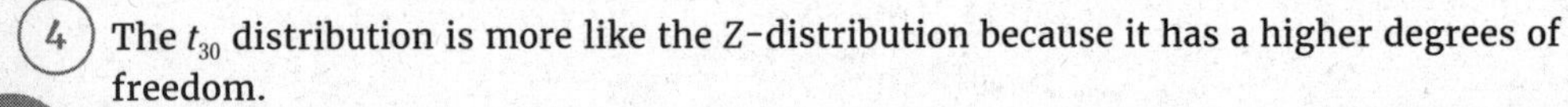

4) The t_{30} distribution is more like the Z-distribution because it has a higher degrees of freedom.

The higher the degrees of freedom, the more the t-distribution looks like a Z-distribution.

5) If t is 2.00 and n is 24, you look at the row for $df=n-1=23$ and follow it across until you see which two numbers 2.00 falls between. It turns out to be between 1.713872 and 2.06866. Go to the top of their columns, and you find that the right-tail probabilities are 0.05 and 0.025. So the right-tail probability for 2.00 is between 0.025 and 0.05.

6) If $t=1.00$ and n is 5, you know that the degrees of freedom are $5-1=4$, so start with row 4 of the t-table. Follow it across and find that $t=1$ lies between 0.740697 and 1.533206. The top of the columns of these two numbers say 0.25 and 0.10, respectively. This means that the right-tail probability for $t=1$ in this case is between 0.10 and 0.25.

7) Because the degrees of freedom are 10 and the right-tail probability is 0.05, you look in row 10 and intersect it with column 0.05 (the right-tail probability) to get $t=1.81$.

8) You have $t=1.96$ with degrees of freedom 30, so go to row 30, and follow across the column until you see 1.697261 and 2.04227. You know 1.96 falls between them, so the right-tail probability lies between their column headings, 0.05 and 0.025, respectively.

9) The value of t is 2, but the sample size is 100, which means the degrees of freedom are 99. That is not one of the rows of the table, so what do you do? Remember that the larger n gets, the closer t gets to Z, so look at the second-to-last row of the t-table, and find the Z-values between which 2.00 falls, 1.95996 and 2.32635. Their columns are 0.025 and 0.01, so the right-tail probability for 2.00 falls between these two values.

If the df of a t-distribution are not on the t-table, you can use the Z-values in the second-to-last row of the t-table to answer your questions.

10) To find the probability that t_9 is less than 0.702722, look in the row for 9 degrees of freedom, and look for 0.702722. The column heading says 0.25, but that is the right-tail, or greater-than, probability, not the less-than probability. However, because the total probability must be 1 for any distribution, the less-than probability is 1 minus the greater-than probability, so the answer is $1-0.25=0.75$.

» Interpreting the central limit theorem

» Using the central limit theorem to find probabilities

» Knowing when you can use the central limit theorem and when you can't

Chapter 9

Demystifying Sampling Distributions and the Central Limit Theorem

Many instructors love to talk on and on about the glories of the central limit theorem (CLT) and how important sampling distributions are to their being, but instructors should all face the fact that you probably don't care about these ideas that much. You just want to get through your class, right? And of course, sampling distributions and any topics related to a "theorem" aren't the easiest subjects on the statistics syllabus (most statistics teaching circles consider them to be the hardest and the most important — what luck). Before you decide to pack it in and call it quits, know that I feel your pain, and I'm here to help.

In this chapter, more than in any other, I think of you as being on a "need-to-know-only" basis. No extra stuff, no frilly theoretical gibberish, no talking about the CLT like it's the best thing since sliced bread, no pining of how "beautiful" all these ideas are. I give you the information that you need to know, when you need to know it, and with plenty of problems to practice. I don't pull any punches here — this stuff is complicated — but I try to break it down so you can focus on only the items you're most likely to face, without the big sales pitch. Can you do it? Yes, you can. Time to get started.

Exactly What Is a Sampling Distribution?

A *sampling distribution* is basically a histogram of all the values that a sample statistic can take on. Sampling distributions are important because when you take a sample from a population, you base all your conclusions on that one sample, and sample results vary. To know how precise your particular sample mean is, you have to think about all the possible sample means you'd get if you took all possible samples of the same size from the population. In other words, to interpret your sample results, you have to know where they stand among the crowd. The crowd of all possible sample statistics you can possibly get is the sampling distribution.

If your data is numerical and your sample statistic is a mean of a sample of size *n*, you should compare that data to all possible sample means of size *n* from that population. You take the sampling distribution for the sample mean, also known as the sampling distribution for $\bar{X}$. A sampling distribution, like any other distribution, has a shape, a center, and a measure of variability. Here are some properties of the sampling distribution for $\bar{X}$:

- If the population already has a normal distribution, the sampling distribution of $\bar{X}$ also has a normal distribution.
- $\bar{X}$ has an approximate normal distribution if the sample size (*n*) is large enough — regardless of what the histogram of the population looks like (due to the central limit theorem [CLT] — the part that instructors get all excited and starry-eyed talking about; see the following section).

  How large is "large enough" to make your calculations work? Statisticians bounce the number $n = 30$ around a lot. The bigger, the better, of course, but the sample size should be at least 30. That means you can take your mean, convert it to a standard (*Z*-) score, and use the *Z*-table to find probabilities for it. (See Chapter 6 for more on standard scores.) All you need to know is what mean and standard deviation to use.
- The mean of all the possible values of $\bar{X}$ is equal to the mean of the population, μ.
- You measure the variability of $\bar{X}$ by the *standard error*. The standard error of a statistic (like the mean) is the standard deviation of all the possible values of the statistic. In other words, the standard error of $\bar{X}$ is the standard deviation of all possible sample means of size *n* you take from the population.

  Here's a nice formula for standard error: The standard error of $\bar{X}$ is equal to the standard deviation of the population divided by the square root of *n*. The notation for this is $\frac{\sigma}{\sqrt{n}}\left(\text{use } \frac{s}{\sqrt{n}} \text{ if } \sigma \text{ is unknown}\right)$.

  As the sample size gets larger, the standard error goes down. A smaller standard error is good because it says that the sample means don't vary by much when the sample sizes get large (your sample mean is precise, as long as your *n* is large and your data was collected properly).

If your sample statistic is a proportion from a sample of size *n*, you should compare it to all possible sample proportions from samples of size *n* from that population. In other words, you find the sampling distribution for the sample proportion, or the sampling distribution for $\hat{p}$. Here are some properties of the sampling distribution for $\hat{p}$:

- The sampling distribution has an approximate normal distribution as long as the sample size is "large enough." (This is due to the CLT; see the following section.)

  How large is "large enough" to make the distribution work? Where *p* is the proportion of the population that have the characteristic of interest, $n * p$ must be at least 10 and $n * (1 - p)$ must be at least 10.

- The mean of all the values of $\hat{p}$ is equal to the original population proportion *p*.
- The standard error of $\hat{p}$ is equal to $\sqrt{\frac{p(1-p)}{n}}$, where *p* is the population proportion. Notice again the formula has a square root of *n* in the denominator, similar to the standard error of $\bar{X}$.
- As the sample size gets larger, the standard error of $\hat{p}$ goes down.

See the following for an example of depicting the sampling distribution for the sample mean.

EXAMPLE

Q. Suppose that you take a sample of 100 from a population that has a normal distribution with mean 50 and standard deviation 15.

a. What sample size condition do you need to check here (if any)?

b. Where's the center of the sampling distribution for $\bar{X}$?

c. What's the standard error?

A. Remember to check the original distribution to see whether it's normal before talking about the sampling distribution of the mean.

a. This sample distribution already has a normal distribution, so you don't need approximations. The distribution of $\bar{X}$ has an exact normal distribution for any sample size. (Don't get so caught up in the $n > 30$ condition that you forget situations where the data has a normal distribution to begin with. In these cases, you don't need to meet any sample size conditions; they hold true for any *n*.)

b. The center is equal to the mean of the population, which is 50 in this case.

c. The standard error is the standard deviation of the population (15) divided by the square root of the sample size (100); in this case, 1.5.

1. Suppose that you take a sample of 100 from a skewed population with mean 50 and standard deviation 15.

   a. What sample size condition do you need to check here (if any)?

   b. What's the shape and center of the sampling distribution for $\bar{X}$?

   c. What's the standard error?

2. Suppose that you take a sample of 100 from a population that contains 45 percent Democrats.

   a. What sample size condition do you need to check here (if any)?

   b. What's the standard error of $\hat{p}$?

   c. Compare the standard errors of $\hat{p}$ for $n = 100$, $n = 1{,}000$, and $n = 10{,}000$ and comment.

Clearing Up the Central Limit Theorem (Once and for All)

The reason that the sampling distributions for $\bar{X}$ and $\hat{p}$ turn out to be normal for large enough samples is because when you're taking averages, everything averages out to the middle. Even if you roll a die with an equal chance of getting a 1, 2, 3, 4, 5, or 6, if you roll it enough times, all the 6s you get average out with all the 1s you get, and that average is 3.5. All the 5s you get average out with all the 2s you get, and that average is 3.5. All the 4s you get average out with all the 3s you get, and that average is (you guessed it) 3.5. Where does the 3.5 come from? From the average of the original population of values 1, 2, 3, 4, 5, 6. And the more times you roll the die, the harder and harder it is to get an average that moves far from 3.5, because averages based on large sample sizes don't change much.

What that means for you is, when you estimate the population mean by using a sample mean (Chapter 11), or when you test a claim about the population mean by using a sample mean (Chapter 13), you can forecast the precision of your results because of the standard error. And what do you need to know to get the standard error of $\bar{X}$? Two things:

- The standard deviation of the population (estimate it with sample standard deviation, *s*, if you don't have it; see Chapter 4)
- The sample size

And you can easily get both numbers from your one single sample. So wow, you can compare your one sample mean to all the other sample means out there, without having to look at any of the others. As Linus would say, "That's what the central limit theorem is all about, Charlie Brown."

TIP

Here's a condensed version of the central limit theorem (CLT) for means that likely resembles any used by instructors. For any population with mean μ and standard deviation σ:

- The distribution of all possible sample means, $\bar{X}$, is approximately normal. (See Chapter 6 for more about normal distribution.)
- The larger the sample size (*n*), the better the normal approximation (most statisticians agree that an *n* of at least 30 does a reasonable job in most cases).
- The mean of the distribution of sample means is also μ.
- The standard error of the sample means is $\frac{\sigma}{\sqrt{n}}$. It decreases as *n* increases.
- If the original data has a normal distribution, the approximation is exact, no matter what the sample size is.

TIP

Here's a condensed version of the theorem applied to proportions. For any population of data with *p* as the overall population percentage:

- The distribution of all possible sample proportions $\hat{p}$ is approximately normal, provided that the sample size is large enough. That is, both $n * p$ and $n * (1 - p)$ must be at least 10. (See Chapter 6 for the normal distribution.)
- The larger the sample size (*n*), the better the normal approximation.
- The mean of the distribution of sample proportions is also *p*.
- The standard error of the sample proportions is $\sqrt{\frac{p(1-p)}{n}}$. It decreases as *n* increases.

See the following for an example of using the CLT with sample proportions.

EXAMPLE

Q. Suppose that you want to find p where p = the proportion of college students in the United States who make more than a million dollars a year. Obviously, p is very small (most likely around 0.1 percent = 0.001), meaning the data is very skewed.

a. How big of a sample size do you need to take to use the CLT to talk about your results?

b. Suppose that you live in a dream world where p equals 0.5. Now what sample size do you need to meet the conditions for the CLT?

c. Explain why you think that skewed data may require a larger sample size than a symmetric data set for the CLT to kick in.

A. Check to be sure the "is it large enough" condition is met before applying the CLT.

a. You need $n*p$ to be at least 10; so, $n*0.001 \geq 10$, which says $n \geq 10 \div 0.001 = 10{,}000$. And checking $n*(1-p)$ gives you $10{,}000(1-.001) = 9{,}990$, which is ≥ 10, which is fine. But wow, the skewness creates a great need for size. You need a larger sample size to meet the conditions with skewed data (since p is far from ½, the case where the distribution is symmetric).

b. You need $n*p$ to be at least 10; so, $n*0.5 \geq 10$, which says $n \geq 10 \div 0.5 = 20$. And checking $n*(1-p)$ gives you $20(1-0.5) = 10$ also, so both conditions check out if n is at least 20. You don't need a very large sample size to meet the conditions with symmetric data ($p = ½$).

c. With a symmetric data set, the data averages out to the mean fairly quickly because the data is balanced on each side of the mean. Skewed data takes longer to average out to the middle, because when sample sizes are small, you're more likely to choose values that fall in the big lump of data, not in the tails.

3 How do you recognize that a statistical problem requires you to use the CLT? Think of one or two clues you can look for. (Assume quantitative data.)

4 Suppose that a coin is fair (so $p = ½$ for heads or tails).

a. Guess how many times you have to flip the coin to get the standard error down to only 1 percent (don't do any calculations).

b. Now use the standard error formula to figure it out.

Finding Probabilities with the Central Limit Theorem

You use $\bar{X}$ to estimate or test the population mean (in the case of numerical data), and you use $\hat{p}$ to estimate or test the population proportion (in the case of categorical data). Because the sampling distributions of $\bar{X}$ and $\hat{p}$ are both approximately normal for large enough sample sizes, you're back into familiar territory somewhat. If you want to find the probability for $\bar{X}$ where X has a normal distribution, all you have to do is convert it to a standard score, look up the value on the Z-table (see the Appendix), and take it from there (and check out Chapter 6 for more on normal distribution).

The only question is how to convert to a standard score. To convert to a standard score, you first find the mean and standard deviation. After you find these figures, you take the number that you want to convert, subtract the mean, and then divide the result by the standard deviation (see Chapter 6 for details). In this chapter, you do everything like you do in Chapter 6, except you replace the standard deviation with the standard error. So in the case of the sample mean, you don't divide by σ; you divide by $\frac{\sigma}{\sqrt{n}}$. Then you look it up on the Z-table and finish the problem from there as usual.

See the following for an example of finding a probability for the sample mean.

EXAMPLE

Q. Suppose that you have a population with mean 50 and standard deviation 10. Select a random sample of 40. What's the chance that the mean will be less than 55?

A. To find this probability, you take the sample mean, 55, and convert it to a standard score, using $z = \frac{\bar{x} - \mu}{\frac{\sigma}{\sqrt{n}}}$. This gives you $\frac{55 - 50}{\frac{10}{\sqrt{40}}} = 3.16$. Using the Z-table, the probability of being less than 3.16 is 0.9993. So the probability that the sample mean is less than 55 is equal to the probability that Z is less than 3.16, which is 0.9992, or 99.92 percent.

5 Suppose that you have a normal population of quiz scores with mean 40 and standard deviation 10.

a. Select a random sample of 40. What's the chance that the mean of the quiz scores won't exceed 45?

b. Select one individual from the population. What's the chance that his/her quiz score won't exceed 45?

6 You assume that the annual incomes for certain workers are normal with a mean of $28,500 and a standard deviation of $2,400.

a. What's the chance that a randomly selected employee makes more than $30,000?

b. What's the chance that 36 randomly selected employees make more than $30,000, on average?

7 Suppose that studies claim that 40 percent of cellphone owners use their phones in the car while driving. What's the chance that more than 425 out of a random sample of 1,000 cellphone owners say they use their phones while driving?

8 What's the chance that a fair coin comes up heads more than 60 times when you toss it 100 times?

When Your Sample's Too Small: Employing the *t*-Distribution

In a case where the sample size is small (and by small, I mean dropping below 30 or so), you have less information on which to base your conclusions about the mean. Another drawback is that you can't rely on the standard normal distribution to compare your results, because the central limit theorem (CLT) can't kick in yet. So what do you do in situations where the sample size isn't big enough to use the standard normal distribution? You use a different distribution — the *t-distribution* (see Chapter 8).

You use the *t*-distribution more when dealing with confidence intervals (see Chapter 11) and hypothesis tests (see Chapter 13). In this chapter, you practice understanding and using the *t*-table (see the Appendix).

See the following for an example of a problem involving averages that use the *t*-distribution.

EXAMPLE

Q. Suppose that you find the mean of 10 quiz scores, convert it to a standard score, and check the table to find out it's equal to the 99th percentile.

a. What's the standard score?

b. Compare the result to the standard score you have to get to be at the 99th percentile on the Z-distribution.

A. *t*-distributions push you farther out to get to the same percentile a Z-distribution would.

a. Your sample size is $n = 10$, so you need the *t*-distribution with $10 - 1 = 9$ degrees of freedom, also known as the t_9 distribution. Using the *t*-table, the value at the 99th percentile is 2.821. (Remember to go to Row 9 of the *t*-table and find the number that falls in the .01 column, since .99 is the probability of being less than the value, and .01 is the greater-than probability — shown in table.)

b. Using the Z-distribution (also in the Appendix), the standard score associated with the 99th percentile is 2.33, which is much smaller than the 2.821 from part a of this question. The number is smaller because the *t*-distribution is flatter than the Z, with more area or probability out in the tails. So to get all the way out to the 99th percentile, you have to go farther out on the *t*-distribution than on the Z.

9. Suppose that the average length of stay in Europe for American tourists is 17 days, with standard deviation 4.5. You choose a random sample of 16 American tourists. The sample of 16 stay an average of 18.5 days or more. What's the chance of that happening?

10. Suppose that a class's test scores have a mean of 80 and standard deviation of 5. You choose 25 students from the class. What's the chance that the group's average test score is more than 82?

11. Suppose that you want to sample expensive computer chips, but you can have only $n = 3$ of them. Should you continue the experiment?

12. Suppose that you collect data on 10 products and check their weights. The average should be 10 ounces, but your sample mean is 9 ounces with standard deviation 2 ounces.

    a. Find the standard score.

    b. What percentile is the standard score found in part a of this question closest to?

    c. Suppose that the mean really is 10 ounces. Do you find these results unusual? Use probabilities to explain.

Answers to Problems in Sampling Distributions and the Central Limit Theorem

1 The fact that you have a skewed population to start with and that you end up using a normal distribution in the end are important parts of the central limit theorem (CLT).

a. The condition is $n > 30$, which you meet.

b. The shape is approximately normal by the CLT. The center is the population mean, 50.

c. The standard error is $\frac{\sigma}{\sqrt{n}} = \frac{15}{\sqrt{100}} = 1.5$. Notice that when the problem gives you the population standard deviation σ, you use it in the formula for standard error. If not, you use the sample standard deviation.

REMEMBER

Checking conditions is becoming more and more of an "in" thing in statistics classes, so be sure you put it on your radar screen. Check conditions before you proceed.

2 Here's a situation where you deal with percents and want a probability. The CLT is your route, provided your conditions check out.

a. You need to check $n * p \geq 10$ and $n * (1-p) \geq 10$. In this case, you have $100 * 0.45 = 45$, which is fine, and $100(1-0.45) = 55$, which is also fine.

b. The standard error is $\sqrt{\frac{p(1-p)}{n}} = \sqrt{\frac{0.45(1-0.45)}{100}} = .050$.

c. The standard errors for $n = 100$, $n = 1{,}000$, $n = 10{,}000$, respectively, are: 0.050, $\sqrt{\frac{0.45(1-0.45)}{1{,}000}} = 0.016$, and $\sqrt{\frac{0.45(1-0.45)}{10{,}000}} = 0.0050$. The standard errors get smaller as n increases, meaning you get more and more precise with your sample proportions as the sample size goes up.

3 The main clue is when you have to find a probability about an average with the data not normal, or a proportion.

TIP

Watch for little words or phrases that can really help you lock on to a problem and know how to work it. I know this sounds cheesy, but there's no better (statistical) feeling than the feeling that comes over you when you recognize how to do a problem. Practicing that skill while the points are free is always better than sweating over the skill when the points cost you something (like on an exam).

4 I'm guessing that you guessed too high in your answer to part a of this question.

a. Whatever you guessed is fine; after all, it's just a guess.

b. The actual answer is 2,500. You can use trial and error and plug different values for n into the standard error formula to see which n gets you a standard error of 1 percent (or 0.01). Or if you don't have that kind of time, you can take the standard error formula, plug in the parts you know, and use algebra to solve for n. Here's how:

$$\sqrt{\frac{p(1-p)}{n}} = 0.01 \rightarrow \sqrt{\frac{0.5(1-0.5)}{n}} = 0.01 \rightarrow \sqrt{\frac{0.25}{n}} = 0.01 \rightarrow \frac{0.25}{n} = (0.01)^2 = 0.0001 \rightarrow$$

$$n(0.0001) = 0.25 \rightarrow n = \frac{0.25}{0.0001} = 2{,}500.$$

TIP

I know what you're thinking: Will my instructor really ask a question like Question 4b? Probably not. But if you train at a little higher setting of the bar, you're more likely to jump over the real setting of the bar in a test situation. If you solved Question 4, great. If not, no big deal.

5 Problems such as this require quite a bit of calculation compared to other types of statistical problems. Hang in there; show all your steps, and you can make it.

TIP

For this problem and the remaining problems in this chapter, I write down every step using probability notation to keep all the information organized and to show you the kind of work your instructor will likely want to see from you when you do these problems. I first write down what I want, in terms of a probability, and then I use the appropriate Z-formula to change the given number to a standard score, and then I look it up on the Z- (or t-) table. Finally, I take 1 minus that value if I'm looking for the probability of being greater than (rather than less than), for example.

a. Not exceeding 45 means < 45, so you want

$$P(\overline{X} < 45) = P\left(Z < \frac{45-40}{\frac{10}{\sqrt{40}}}\right) = P(Z < 3.16) = 0.9992, \text{ or } 99.92\%$$

by looking it up on the Z-table (see the Appendix).

b. In this case, you focus on one individual, not the average, so you don't use sampling distributions to answer it; you use the old Z-formula (see Chapter 6). That means you want $P(\overline{X} < 45)$. Take 45 and convert it to a Z-score (subtract the mean, 40, and divide by the standard deviation, 10) to get $Z = 0.5$. The probability of being less than 0.5 using the Z-table is 0.6915, or 69.15%.

TIP

Be on the lookout for two-part problems where, in one part, you find a probability about a sample mean, and in the other part, you find the probability about a single individual. Both convert to a Z-score and use the Z-table in the Appendix, but the difference is the first part requires you to divide by the standard error, and the second part requires you to divide by the standard deviation. Instructors really want you to understand these ideas, and they put them on exams almost without exception.

6 Here's another problem that dedicates one part to a probability about one individual (part a) and another part to a sample of individuals (part b).

a. You have $P(X > 30{,}000) = P\left(Z > \frac{30{,}000 - 28{,}500}{2{,}400}\right) = P(Z > 0.63) = 1 - 0.7357 = 0.2643$. You get this answer by changing 30,000 to a Z-score of 0.63, looking up 0.63 on the Z-table, and taking 1 minus the result because you have a greater-than probability.

b. Here, you want

$$P(\overline{X} > 30{,}000) = P\left(Z > \frac{30{,}000 - 28{,}500}{\frac{2{,}400}{\sqrt{36}}}\right) = P\left(Z > \frac{1{,}500}{400}\right) = P(Z > 3.75) = 1 - P(Z > 3.75).$$

The Z-value of 3.75 is pretty much off the chart when you look at the Z-table. If this happens to you, use the last value on the chart (in this case, 3.69) and say that the probability of being beyond 3.75 on the Z-table has to be smaller than the probability of being beyond 3.69, the last value on the chart. The percentile for 3.69 is 99.99 percent, so the area beyond (above) that is $100\% - 99.99\% = 0.01\%$ (or 0.0001). Therefore, you can say that the probability of 36 workers making more than $30,000 is less than 0.0001.

7. Here you have to find the sample proportion by using the information in the problem. Because 425 people out of the sample of 1,000 say they use cellphones while driving, you take 425 divided by 1,000 to get your sample proportion, which is 0.425. Now you want to know how likely it is to get results like that (or greater than that). That means you want

$$P\left(\hat{p} > \frac{425}{1,000}\right) = P(\hat{p} > 0.425) = P\left(Z > \frac{0.425 - 0.4}{\sqrt{\frac{0.4(1-0.4)}{1,000}}}\right) = P\left(Z > \frac{0.025}{0.015}\right) = P(Z > 1.67) = 1 - 0.9525 = 0.0475$$

or 4.75%.

TIP

If you're given the sample size and the number of individuals in the group you're interested in, divide those to get your sample proportion.

8. A fair coin means $p = 0.5$, where p is the proportion of heads or tails in the population of all possible tosses. In this case, you want the probability that your sample proportion is beyond $60 \div 100 = 0.60$. So you have $P(\hat{p} > 0.60) = P\left(Z > \frac{0.6 - 0.5}{\sqrt{\frac{0.5(1-0.5)}{100}}}\right) = P(Z > 2.04) = 1 - 0.9793 = 0.0207$ or 2.07%.

9. You have $P\left(t_{15} > \frac{18.5 - 17}{\frac{4.5}{\sqrt{16}}}\right) = P(t_{15} > 1.33) = 1 - P(t_{15} < 1.33) = 0.10$ or 10%.

First, change 18.5 to a value on the t-distribution (1.33). Then look at the t-table (Appendix) in the row for $16 - 1 = 15$ degrees of freedom, and find the number closest to 1.33 (which is 1.341). This gives you the answer 0.10.

10. You want $P\left(t_{24} > \frac{82 - 80}{\frac{5}{\sqrt{25}}}\right) = P\left(t_{24} > \frac{2}{1}\right) = P(t_{24} > 2) = 0.025$ or 2.5%.

First, change the 82 to a value on the t-distribution (2), and then look at the t-table in the row for $25 - 1 = 24$ degrees of freedom, and find the number closest to 2 (which is 2.064). This gives you the answer 0.025.

11 The experiment might not be worth it because the values are so large on the t-distribution with two degrees of freedom; you have to deal with too much variability in what you expect to find.

12 Here you compare what you expect to see with what you actually get (which comes up in hypothesis testing; see Chapter 13). The basic information here is that you have $\bar{x} = 9$, $s = 2$, $\mu = 10$, and $n = 10$.

a. The standard score is $\frac{\bar{x}-\mu}{s/\sqrt{n}} = \frac{9-10}{2/\sqrt{10}} = \frac{-1}{0.632} = -1.58.$

Note: This number is negative, meaning you're 1.58 standard deviations below the mean on the t_9 distribution ($n-1 = 10-1 = 9$). There are no negative values on the t-table. What you need to do is take 100 percent minus the percentile you get for the positive value 1.58. Now 1.58 lies between the 90th and 95th percentiles on the t_9 distribution, so -1.58 lies between the $100-90 = 10$th percentile and the $100-95 = 5$th percentile on the t_9 distribution.

TIP

To get percentiles for negative standard scores on the t-table, take 100 percent minus the percentile for the positive version of the standard score. Because the t-distribution is symmetric, the area below the negative standard score is equal to the area above the positive version of the standard score. And the area above the positive standard score is 100 percent minus what's found on the t-table.

b. Your standard score of 1.58 from part a of this question is between the 90th and 95th percentiles on the t_9 distribution (that is, between 1.383 and 1.833) — a little closer to the 90th. That's the best answer you can give using the t-table.

c. The results aren't entirely unusual because, according to part b of this question, they happen between 5 and 10 percent of the time.

3 Guesstimating and Hypothesizing with Confidence

IN THIS PART . . .

Review the basics of confidence intervals and hypothesis tests.

Follow detailed steps on how to set up, carry out, and interpret confidence intervals and hypothesis tests for one and two means and one and two proportions.

» Knowing how the components affect the size of the margin

» Computing the margin of error for means and proportions

» Accurately interpreting margin of error

Chapter 10
Making Sense of Margin of Error

Margin of error is a critical element for understanding confidence intervals, a heavyweight topic on introductory statistics syllabi. Although the calculations are very important, you also need to look at the components of margin of error and how they each play a role in what the value of the margin of error is. Another important element is knowing what margin of error actually measures. Looks can be deceiving. This chapter goes over the basics of margin of error; Chapter 11 focuses on actually calculating confidence intervals, based on these margin-of-error ideas, and Chapter 12 deals with interpreting your confidence intervals properly.

Reviewing Margin of Error

One of the main caveats of sample results is that they vary from sample to sample, so you shouldn't report statistical results without also including the *margin of error.* The margin of error is the amount by which you expect the sample results to vary from sample to sample. A large margin of error means you expect the sample results to change a lot (not a good thing), and a small margin of error means you don't expect the sample results to change much with repeated sampling (a good thing).

In introductory statistics, you tackle problems involving the margin of error for the sample mean or the sample proportion. If your data is quantitative and you want to determine the average value for the population, you use the sample mean, plus or minus the margin of error for your sample mean. If your data is categorical and you want to determine the percentage in the population with a certain characteristic, you use the sample proportion, plus or minus the margin of error for your sample proportion.

The margin of error formulas look different on the surface in each situation, but they each contain three important components:

- **The Z^* value:** A number from the standard normal (Z) distribution (see Chapter 6). The Z^* value tells you the number of standard deviations (or standard errors) you should add and subtract to get the level of confidence you want. (Use the Z when your sample size is large. For small samples, you use the *t*-distribution rather than the Z; see Chapter 8.)

  TIP

  In Chapter 11, you get more information about Z^* values, but for the problems in this chapter, the following Z^* values will get you through. For 80 percent confidence, use $Z^* = 1.28$; for 90 percent confidence, use $Z^* = 1.64$; for 95 percent confidence, Z^* is 1.96; and for 99 percent confidence, Z^* is 2.58. (Note that each Z^* gets larger as the level of confidence goes up.) Keep these Z^* values handy as you do the problems in this chapter.

  Textbooks will differ in terms of the exact number of standard deviations to add/subtract for 90 percent confidence. The most accurate answer is exactly between 1.64 and 1.65, which is 1.645. Throughout this workbook, however, I use 1.64. Ask your professor which one he or she prefers.

- **The standard deviation of the population:** The amount of variability that exists in the population in terms of the variable you're measuring. If the standard deviation isn't available for the population, you substitute the standard deviation of the sample and use the *t*-distribution (more in Chapter 11).

- **The sample size:** The total number of values in the data set — denoted *n*.

The general formula for margin of error (if n is large) is: $\pm Z^*$ times the standard deviation divided by the square root of *n* (where *n* is the sample size). You generally take your sample statistic (the mean or proportion) and add or subtract the margin of error to get a confidence interval for the population value. (More about generating confidence intervals in Chapter 11.) See the following for an example of the importance of margin of error.

EXAMPLE

Q. You hear in a dentist's ad that only 45 percent of adults floss their teeth daily. Explain why these results are virtually meaningless without the margin of error.

A. Without a margin of error, you have no way of knowing how precise or consistent these results would be from sample to sample. They can vary so much that the 45 percent becomes unrecognizable, or they can vary so little that the 45 percent is about as close as you can get to the truth.

1. Suppose that you want to be 99 percent confident in your results, and you plan on having a large sample size. What's your Z^* value?

2. Suppose that polling for the upcoming election says the results are "too close to call," because they fall within the margin of error. Explain what the pollsters mean.

3. Explain why it makes sense that the margin of error should depend on the standard deviation of the population.

4. What margin of error should you require to have 100 percent confidence in your sample result?

Calculating the Margin of Error for Means and Proportions

The specific formula for margin of error for the sample mean is $\pm Z^* * \frac{\sigma}{\sqrt{n}}$.

The specific formula for margin of error for the sample proportion is $\pm Z^* * \sqrt{\frac{\hat{p}(1-\hat{p})}{n}}$.

Notice that each formula contains the three components I mention earlier in this chapter. In the case of the sample proportion, because the data is categorical (yes/no or success/failure) data, the standard deviation is the square root of the estimate of p (the proportion of yes answers in the population) times the estimate of $1-p$ (the proportion of noes in the population). In this case, the estimate of the proportion of yeses in the population is the proportion of yeses in the sample — hence, the carrot or "hat" sign over the p (to be read as "p-hat"). The estimate of the proportion of noes in the population is the proportion of noes in the sample (read as "one minus p-hat"). See the following for an example of whether to use margin of error for proportions or means.

EXAMPLE

Q. Suppose that you want to estimate the proportion of cellphone users at a university. Which formula for margin of error do you need?

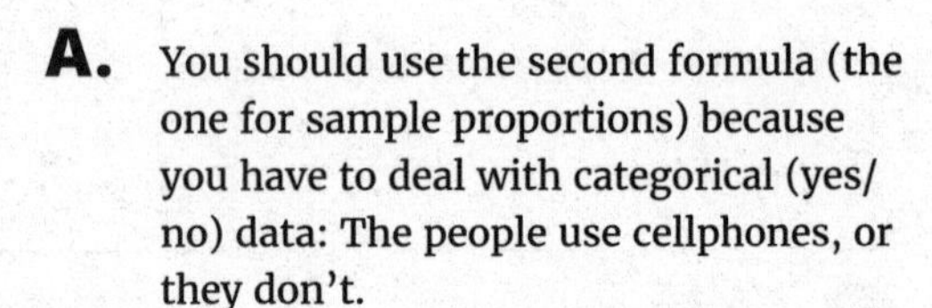

A. You should use the second formula (the one for sample proportions) because you have to deal with categorical (yes/no) data: The people use cellphones, or they don't.

5 A survey of 1,000 dental patients produces 450 people who floss their teeth adequately. What's the margin of error for this result? Assume 90 percent confidence.

6 A survey of 1,000 dental patients shows that the average cost of a regular six-month cleaning/checkup is $150.00 with a standard deviation of $80. What's the margin of error for this result? Assume 95 percent confidence.

7 Out of a sample of 200 babysitters, 70 percent are girls, and 30 percent are guys.

a. What's the margin of error for the percentage of female babysitters? Assume 95 percent confidence.

b. What's the margin of error for the percentage of male babysitters? Assume 95 percent confidence.

8 You sample 100 fish in Pond A at the fish hatchery and find that they average 5.5 inches with a standard deviation of 1 inch. Your sample of 100 fish from Pond B has the same mean, but the standard deviation is 2 inches. How do the margins of error compare? (Assume the confidence levels are the same.)

9 Suppose that you conduct a study twice, and the second time you use four times as many people as you did the first time. How does the change affect your margin of error? (Assume the other components remain constant.)

10 Suppose that Sue and Bill each make a confidence interval out of the same data set, but Sue wants a confidence level of 80 percent compared to Bill's 90 percent. How do their margins of error compare?

11 Suppose that you find the margin of error for a sample proportion. What unit of measurement is it in?

12 Suppose that you find the margin of error for a sample mean. What unit of measurement is it in?

Increasing and Decreasing Margin of Error

Because the margin of error for sample means and proportions contains three components, you can imagine that changing the values of each component has some effect on the margin of error. People often look at the different possibilities for margin of error before they start their study. They do so to get a feeling for how big their sample size should be or how much confidence they can have in their results, given the sample size they can afford or the amount of variability they attribute to their population. Knowing how changes in each component affect the margin of error should be an important item on your statistical checklist. See the following for an example of how sample size affects margin of error.

EXAMPLE

Q. Suppose that you increase a sample size and keep everything else the same. What happens to the margin of error?

A. The margin of error decreases, because n is in the denominator of a fraction, and increasing the denominator decreases the fraction. This also makes sense because including more data should increase the level of precision in your results (provided you include good data).

13 What happens to the margin of error if you increase your confidence level (but keep all other elements fixed)?

14 Suppose that you have two ponds of fish in a fish hatchery, and the first pond has twice as much variability in fish lengths as the second. You take a sample of 100 fish from each pond and calculate 95 percent confidence intervals for the average fish lengths for each pond. Which confidence interval has a larger margin of error?

15 How can you increase your confidence level and keep the margin of error small? (Assume you can do anything you want with the components of the confidence interval.)

16 Suppose that you conduct a pilot study and find that your target population has a very large amount of variability. What can you do (if anything) to ensure a small margin of error in your full-blown study?

Interpreting Margin of Error Correctly

Margin of error measures the level of precision in your sample results. Precision represents consistency, but it doesn't necessarily promise on-target results (they may be systematically over or under by a certain amount each time). In other words, margin of error doesn't measure the amount of bias that may come with the sample results. Most types of bias are extremely difficult to measure, so you should do your best to minimize bias wherever possible.

EXAMPLE

Q. An Internet survey has 50,000 respondents, and the margin of error is plus or minus 0.1 percent. Do you believe the results are that precise? Explain.

A. No. The results are based on an Internet survey where people select themselves to participate. Therefore, the reported margin of error isn't meaningful, because the error in the sample results is off by way more than that amount. The survey is biased because you find it on the Internet. So even though the numbers you plug in to the margin of error formula seem nice and small, they're based on garbage and should be ignored.

17 You find out that the dietary scale you use each day is off by a factor of 2 ounces (over — at least that's what you say!). The margin of error for your scale was plus or minus 0.5 ounces before you found this out. What's the margin of error now?

18 You're fed up with keeping Fido locked inside, so you conduct a mail survey to find out people's opinions on the new dog barking ordinance in a certain city. Of the 10,000 people who receive surveys, 1,000 respond, and only 80 are in favor of it. You calculate the margin of error to be 1.2 percent. Explain why this reported margin of error is misleading.

19 Does the margin of error measure the amount of error that goes into collecting and recording data?

20 If you add and subtract the margin of error to/from the sample mean, do you guarantee the population mean to be in your resulting interval?

21 A television news channel samples 25 gas stations from its local area and uses the results to estimate the average gas price for the state. What's wrong with its margin of error?

22 If you control for all possible confounding variables in an experiment, can you reduce the margin of error in your results?

Answers to Problems in Making Sense of Margin of Error

1. The Z^* value for a 99 percent confidence interval is 2.58, because the area on the Z-distribution between –2.58 and +2.58 is about 0.99, or 99 percent.

2. They mean that after you add and subtract the margin of error from one of the percentages in the poll, the other percentage is included in that interval, so they can't be statistically different. Here's another way to look at this: If 49 percent of responders say they want to vote for Candidate A, and 51 percent back Candidate B, with a margin of error of plus or minus three points, Candidate A may get anywhere from 46 percent to 52 percent of the vote in the population. Candidate B may get anywhere from 48 percent to 54 percent of the vote. The intervals overlap, which means the results are too close to call.

3. If a population has a wide amount of variety in its values, its average value becomes harder to pinpoint, which makes the margin of error larger. You can offset this, however, by sampling more data, because that increases *n* and offsets the larger standard deviation in the margin of error formula.

4. You need an infinite margin of error in the case of the sample mean, and at least a 50 percent margin in the case of the sample proportion (because you want to cover all possible values). Having 100 percent confidence is meaningless, however. Who wants to say that they know the percentage of people owning cellphones is 0 to 100 percent?

5. This problem concerns the percentage of people in a certain category (those who floss their teeth daily). The formula to use is the margin of error for the sample proportion. Note that Z^* is 1.64 because the confidence level is 90 percent, and that $\hat{p}$, the sample proportion, is $450 \div 1{,}000 = .45$.

   The margin of error is $\pm Z^* * \sqrt{\frac{\hat{p}(1-\hat{p})}{n}} = \pm 1.64\sqrt{\frac{0.45(1-0.45)}{1{,}000}} = \pm 1.64(.0157) = \pm .026$ or 2.6%.

REMEMBER

Any formula involving the sample proportion, $\hat{p}$, requires that you use the decimal version of the percentage (also known as the proportion), or you get inaccurate results. For example, if the sample proportion is 70 percent, you must use 0.70 for $\hat{p}$ in the formula.

6. This problem concerns the average of a quantitative variable (cost of dental cleaning and exam). The formula to use is the margin of error for the sample mean. Note that Z^* is 1.96, because the confidence level is 95 percent. The margin of error is

   $\pm Z^* * \frac{\sigma}{\sqrt{n}} = \pm 1.96\frac{80}{\sqrt{1{,}000}} = \pm 1.96(2.53) = \pm 4.96$ (dollars).

TIP

When making margin of error calculations, keep at least two significant digits after the decimal point throughout the calculations, rounding only at the very end to avoid accumulated round-off errors.

7. This problem concerns the percentage of babysitters in a certain category, so you have to use the formula for the margin of error for the sample proportion.

a. Here you concentrate on the percentage of female babysitters. Note that Z^* is 1.96, because the confidence level is 95 percent, and that $\hat{p}$, the sample proportion, is 0.70. The margin of error is $\pm Z^* * \sqrt{\frac{\hat{p}(1-\hat{p})}{n}} = \pm 1.96\sqrt{\frac{0.70(1-0.70)}{200}} = \pm 1.96(0.0324) = \pm 0.0635$.

b. Here you concentrate on the percentage of male babysitters. Note that Z^* is 1.96, because the confidence level is 95 percent, and that $\hat{p}$, the sample proportion, is 0.30. The margin of error is $\pm Z^* * \sqrt{\frac{\hat{p}(1-\hat{p})}{n}} = \pm 1.96\sqrt{\frac{0.30(1-0.30)}{200}} = \pm 1.96(0.0324) = \pm 0.0635$. Notice the similarities to answer a, because 0.30 equals $1-.70$, and 0.70 equals $1-.30$, so you plug the exact same numbers into the equation.

8. The sample from Pond B has a larger margin of error because the standard deviation is larger, and standard deviation is involved in the numerator of the fraction for margin of error.

9. It reduces the margin of error by a factor of 2, because 2 is the square root of 4. (You substitute n in the margin of error equation with $4n$, and the square root of $4n$ is 2 times the square root of n.) Notice: The change doesn't reduce margin of error by a factor of 4, because n is under a square root sign.

   To cut the margin of error in half, quadruple the sample size.

10. This problem looks at the effect of confidence level on the size of the margin of error. Sue's margin of error is smaller than Bill's, because Sue's Z^* value is 1.28, and Bill's is 1.64.

REMEMBER

A larger Z^* value makes the margin of error larger (if the other components remain the same).

11. The margin of error for a sample proportion is in the same units as the proportion — a number between 0 and 1. You can rewrite it in the end as a percentage if you want, after the calculations are done.

12. The margin of error for a sample mean is in the same units as the original data. For example, if you want to calculate the margin of error for average fish length, and the fish are measured in inches, the margin of error is in inches as well.

TIP

Keep track of the units you work with when you do your calculations. This helps you recognize when something just doesn't appear right, and it scores you big points with your professor if you write down the (correct) units — trust me!

13. Margin of error increases if you increase the confidence level and keep all the other components the same, because as your confidence level increases, the Z^* value increases.

14. The first pond has more variability in the population, so the value of the standard deviation is larger. Because standard deviation appears in the top part of the fraction in margin of error, increasing the standard deviation increases the margin of error. This makes good sense, because more variability in the population makes it harder to pin down the actual population average with precision.

TIP

The population standard deviation for quantitative data is denoted by σ. You typically don't know it, so you substitute s, the standard deviation of the sample, in the formulas, but only if the sample size is large enough (typically more than 30). If not, you use a t-distribution value (see Chapter 7) in place of the Z^* value.

15. If you increase the confidence level, you increase the Z^* value and thereby increase the margin of error. You can offset this increase by also increasing the sample size, and because the sample size is in the denominator of margin of error, increasing it has the opposite effect — lowering the margin of error. This makes good sense, because having more data allows you to be more precise.

REMEMBER

You can offset an increase in margin of error by increasing the sample size.

16. Increase the sample size in your actual (full-blown) study to offset the anticipated increase in margin of error that stems from the larger value of s.

17. The margin of error doesn't change, because it doesn't depend on what the average measurements are; it depends on the standard deviation and sample sizes. Also, margin of error doesn't take bias into account, and this scale is a biased scale.

REMEMBER

Margin of error measures precision (consistency), but it doesn't measure bias (being systematically over or systematically under the true value).

18. This survey is biased because of the very low number of people who responded. The response rate for this survey (number of respondents divided by total number sent out) is only $1{,}000 \div 10{,}000 = 0.10$ or 10%. That means 90% of the people who received the survey didn't respond. If you base the results only on those who responded, you would say $80 \div 1{,}000 = 0.08$ or 8% of them are in favor of the dog barking ordinance, and the other 92% are against it. This may lead you to believe that the results are biased, and indeed they are. Those who have the strongest opinions respond to surveys, in general. Suppose that the 9,000 people who didn't respond were actually in favor of the dog ordinance. That would mean a total of $9{,}000 + 80 = 9{,}080$ in favor (and $9{,}080 \div 10{,}000 = 0.908$ or 90.8%) and 920 against (that percentage would be $920 \div 10{,}000 = 0.092$ or 9.2%). Those percentages change dramatically from the results based only on the respondents.

    Bottom line: Respondents and nonrespondents are not alike and should not be assumed to be so. If the response rate of a survey is too low, you should ignore the results; they are likely to be biased toward those with strong opinions.

WARNING

Don't be misled by seemingly precise studies that offer a small margin of error without looking at the quality of the data. If the data is bad, the results are garbage, even if the formulas don't know that.

19. No. Margin of error measures the sampling error only, which is the error due to the random sampling process. The word "error" makes it seem like human error, but it actually means random error (error due to chance alone).

20. No, you can never be 100 percent confident, unless you include all possible values in your interval, which renders it useless. The confidence level tells you what percentage of the samples you expect to yield correct intervals. For example, if your confidence level is 95 percent, you can expect that 95 percent of the time your sample produces an interval that contains the true population value, and that 5 percent of the time it doesn't, just by random chance.

21. The margin of error is meaningless because the station bases the sample on local area gas stations that don't represent a statewide sample.

22. Possibly, only because the control may reduce the amount of variability in the results (as well as eliminate bias, which doesn't affect margin of error). By controlling for certain variables, you can reduce the variability.

» Calculating confidence intervals for one and two means and proportions

» Knowing which confidence interval to use

» Interpreting confidence intervals correctly

Chapter 11 Calculating Confidence Intervals

Introductory statistics looks at *confidence intervals* for means and proportions, from one or two populations. You use a confidence interval to find the population parameter (the population mean, the proportion in the population with a certain characteristic, and so on). In other words, you want to make a good guess, or a good estimate, as to what that population value is. You don't use a confidence interval if you already have a claim or idea about what the population value is to test that claim. That situation calls for a *hypothesis test* (see Chapter 13).

Walking through a Confidence Interval

All confidence intervals contain the same basic parts: a sample statistic, plus or minus a margin of error. The *margin of error* measures how much you expect the sample statistic to vary from one sample to the next (see Chapter 10). The formulas for margin of error that you see in a statistics course all involve the same idea: a Z^* value times the standard error. The Z^* value is a number from the standard normal (Z-) distribution and reflects the number of standard errors you need to add/subtract to get the confidence level you want. For example, to be 95 percent confident, you need to add/subtract about two standard errors (1.96 to be exact) because of the empirical rule (see Chapter 2 for more information). To be more confident, add/subtract more standard errors.

There are four basic steps to calculating any confidence interval:

1. **Find your sample estimate.**
2. **Calculate the margin of error.**
3. **Take the sample estimate, plus and minus the margin of error to get a range, or interval.**
4. **Interpret your results by talking about how confident you are that the population parameter is actually in your interval.**

Interpreting a confidence interval can be tricky. You can't say that a 95 percent confidence interval means there is a 95 percent chance that the population parameter is in the particular interval that you calculate (although it certainly seems that way!). Most people say a 95% confidence interval means they are 95% confident that the parameter is in the interval. It means that of all the times you apply a 95 percent confidence interval to a new sample of data, 95 percent of the time you're right — the population parameter is in the interval. The other 5 percent of the time, you get an incorrect answer, just by chance. This 5 percent is an example of this chance error of being incorrect, and is denoted in general by the letter α. The amount of confidence you have in your results is called the *confidence level* (in this example, the 95 percent) and is denoted by $1-\alpha$ (alpha is the chance of making an error; more on this in Chapter 12). Table 11-1 shows common confidence levels and their corresponding Z^* values.

Also, stating your statistical conclusions (giving the numbers) isn't really an interpretation of those results, so don't stop there. Tell what those results mean in practical terms; explain what the results mean to the person conducting the study.

Table 11-1 Common Confidence Levels and the Matching Z^* Values

Confidence Level	Z^* Value
80%	1.28
90%	1.64
95%	1.96
98%	2.33
99%	2.58
99.7%	2.96

See the following for an example of identifying the various parts of a confidence interval.

Q. Suppose that you're trying to estimate the population mean, and your sample size of 100 gives you a mean of 9, and your margin of error is 3 for a 95 percent confidence interval.

a. What is the resulting confidence interval?

b. What is the confidence level?

c. What is alpha?

A. **You need to be able to pick out all the various parts of a confidence interval, even before you can calculate one for yourself.**

a. The resulting confidence interval here is 9 ± 3 because 3 is the margin of error. Another way to write this interval is (6, 12) because $9+3=12$ and $9-3=6$. *Interpretation:* You're 95 percent confident that the population mean is in the interval (6, 12).

b. The confidence level is 95 percent (given in the problem).

c. Alpha is $100 \text{ percent} - 95 \text{ percent} = 5 \text{ percent}$, or 0.05.

---

1 Suppose that a 95 percent confidence interval for the average number of minutes a regular (obviously not teenage) customer uses a cellphone in a month is 110 plus or minus 35 minutes.

a. What's the margin of error?

b. What are the lower and upper boundaries for this confidence interval?

Suppose that you make two confidence intervals with the same data set — one with a 95 percent confidence level and the other with a 99.7 percent confidence level.

a. Which interval is wider?

b. Is a wide confidence interval a good thing?

3 Is it true that a 95 percent confidence interval means you're 95 percent confident that the sample statistic is in the interval?

4 Is it true that a 95 percent confidence interval means there's a 95 percent chance that the population parameter is in the interval?

Deriving a Confidence Interval for a Population Mean

You use a confidence interval for a population mean under two conditions:

- The variable you deal with is quantitative, such as height, weight, IQ, length, or test score.
- You want to estimate the average or mean value of this variable among the population.

If your population is normal or the sample size is large enough (more than 30), the formula for a $(1-\alpha)$ percent confidence interval for the population mean is $\bar{x} \pm Z^{*} * \frac{\sigma}{\sqrt{n}}$.

You need four pieces of information to calculate this confidence interval: the sample mean $\bar{x}$, the standard deviation, the sample size (n), and the confidence level (95 percent, 99 percent, and so on). If the population standard deviation, σ, is known, use it in the formula. If not, use the sample standard deviation (s).

To calculate the confidence interval:

1. **Calculate the margin of error and write down your result.**

   The margin of error is the part that comes after the ± sign.

2. **Take your sample mean and subtract the margin of error to get the lower bound on the confidence interval.**
3. **Take the sample mean and add the margin of error to get the upper bound.**
4. **Last but not least, interpret your results.**

TIP

Most instructors will make a big deal of completing the last step. They also may prefer you to actually add and subtract the margin of error (Steps 2 and 3) so you have an actual interval for your answer, not just a number with a plus or minus directive.

For small samples, or if the population standard deviation is unknown, you use the sample standard deviation, s, and a t^* value (see the last row of the t- table in the Appendix) rather than a Z^* value (from Table 11-1) in your margin of error calculations.

See the following for an example of calculating a confidence interval for a population mean.

EXAMPLE

Q. Suppose that 100 randomly selected used cars on a lot have an average of 30,250 miles on them, with a population standard deviation of 500 miles. Find a 95 percent confidence interval for the average miles on all the cars in this lot.

A. You need to find a 95 percent confidence interval for the population mean, and the formula you need is $\bar{x} \pm Z^* * \frac{\sigma}{\sqrt{n}}$. The sample mean, $\bar{x}$, is 30,250; the population standard deviation, σ, is 500; and Z^* is 1.96 because the confidence level is 95 percent (see Table 11-1). So your confidence interval is $\bar{x} \pm Z^* * \frac{\sigma}{\sqrt{n}} = 30{,}250 \pm 1.96 * \frac{500}{\sqrt{100}} =$ $30{,}250 \pm 98 = (30{,}152, 30{,}348)$ miles. *Interpretation:* You're 95 percent confident that the average mileage for all cars in this lot is between 30,152 and 30,348.

5 Tines can range from 2 to upwards of 50 or more on a male deer. You want to estimate the average number of tines on the antlers of male deer in a nearby metro park. A sample of 30 deer has an average of 5 tines, with a population standard deviation of 3.

a. Find a 95 percent confidence interval for the average number of tines for all male deer in this metro park.

b. Find a 98 percent confidence interval for the average number of tines for all male deer in this metro park.

6 Based on a sample of 100 participants, a 95 percent confidence interval for the average weight loss the first month under a new weight-loss plan is 11.4 pounds, plus or minus 0.51.

a. Explain what this confidence interval means.

b. What's the margin of error for this confidence interval?

7 A 95 percent confidence interval for the average miles per gallon for all cars of a certain type is 32.1, plus or minus 1.8. The interval is based on a sample of 40 randomly selected cars.

a. What units represent the margin of error?

b. Suppose that you want to decrease the margin of error, but you want to keep 95 percent confidence. What should you do?

8 Suppose that you want to increase the confidence level of a particular confidence interval from 80 percent to 95 percent without changing the width of the confidence interval. Can you do it?

Figuring a Confidence Interval for a Population Proportion

You use a confidence interval for a population proportion under three conditions:

- The variable you deal with is categorical, such as gender, political affiliation, opinion (agree, disagree, no opinion), or marital status.
- You want to estimate the proportion in the population who fall into one of the specific categories by using a random sample.
- The sample size is "large enough."

The formula for a $(1-\alpha)$ percent confidence interval for the population proportion is

$$\hat{p} \pm Z^* * \sqrt{\frac{\hat{p}(1-\hat{p})}{n}}.$$

You need three pieces of information to calculate this confidence interval:

- The sample proportion, $\hat{p}$
- The sample size (n)
- The confidence level (95 percent, 99 percent, and so on)

The sample proportion is the proportion (or percent) of individuals in the sample who fall under the category of interest (for example, proportion of males in the sample). You find the sample proportion by taking the total number of individuals in the sample who have that characteristic divided by the sample size (n). The sample size is large enough if $n * \hat{p}$ is at least 10 and $n * (1 - \hat{p})$ is at least 10. This occurs in the vast majority of survey situations (where you typically use this confidence interval).

See the following for an example of calculating a confidence interval for a population proportion.

EXAMPLE

Q. A random sample of 1,000 U.S. college students finds that 28 percent watch the Super Bowl every year. Find a 95 percent confidence interval for the proportion of all U.S. college students who watch the Super Bowl every year.

A. The sample proportion is 0.28, $n = 1{,}000$, and $Z^* = 1.96$ (from Table 11-1), so the 95 percent confidence interval is

$0.28 \pm 1.96 * \sqrt{\frac{0.28(1-0.28)}{1{,}000}} =$

$0.28 \pm 1.96 * \sqrt{0.0002} = 0.28 \pm .028 =$ (0.252, 0.308). *Interpretation:* You're 95 percent confident that the proportion of all U.S. college students who watch the Super Bowl every year is between 0.252 and 0.308, or 25.2 percent and 30.8 percent.

---

9 A random sample of 1,117 U.S. college students finds that 729 go home at least once each term. Find a 98 percent confidence interval for the proportion of all U.S. college students who go home at least once each term.

10 A poll of 2,500 people shows that 50 percent approve of a smoking ban in bars and restaurants. What's the margin of error for this confidence interval? (Assume 95 percent confidence.)

11 Suppose that 73 percent of a sample of 1,000 U.S. college students drive a used car as opposed to a new car or no car at all.

a. Find an 80 percent confidence interval for the percentage of all U.S. college students who drive a used car.

b. What sample size would cut this margin of error in half?

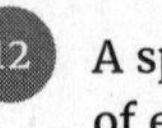

12 A special interest group reports a tiny margin of error (plus or minus 0.04 percent) for its online survey based on 50,000 responses. Is the margin of error legitimate? (Assume that the group's math is correct.)

Calculating a Confidence Interval for the Difference of Two Means

You use a confidence interval for the difference of two population means under three conditions:

- The variable you deal with is quantitative, such as height, weight, or exam score.
- You want to compare the means of two independent (separate) groups or populations.
- Both sample sizes are large enough (more than 30).

The formula for a $(1-\alpha)$ percent confidence interval for the difference of two means when both populations are normal or approximately normal and the population standard deviations are known is $\left(\bar{x}-\bar{y}\right)\pm Z^{*}*\sqrt{\frac{\sigma_1^2}{n_1}+\frac{\sigma_2^2}{n_2}}$.

Because you want to compare two separate (independent) populations, you need two separate (independent) samples — one from each population. Besides the usual Z^{*} value, to calculate this confidence interval, you need the sample mean, population standard deviation, and sample size for the first population (indicated by $\bar{x}$, σ_1, and n_1, respectively) and the sample mean, population standard deviation, and sample size for the second population (indicated by $\bar{y}$, σ_2, and n_2, respectively). Subtract the mean of the first group minus the mean of the second group to get your sample statistic (for the difference). Add and subtract the margin of error and interpret the results.

TIP

If the populations are not normal and the sample sizes are small and/or the population standard deviations are unknown, you substitute the sample standard deviations s_1 and s_2 for the population standard deviations, and your test statistic is $(\bar{x}-\bar{y}) \pm t^* * \sqrt{\frac{s_1^2}{n_1}+\frac{s_2^2}{n_2}}$ where t^* comes from the bottom row of the t-table (see the Appendix), in the row for $n_1 + n_2 - 2$ degrees of freedom (see Chapter 8).

See the following for an example of finding a confidence interval for the difference between two population means.

EXAMPLE

Q. Suppose that 100 randomly selected used cars on Lot 1 have an average of 35,328 miles on them, with a standard deviation of 750 miles. A hundred randomly selected used cars on Lot 2 have an average of 30,250 miles on them, with a standard deviation of 500 miles. Find a 95 percent confidence interval for the difference in average miles between all cars on Lots 1 and 2.

A. Using Lot 1 as group one and Lot 2 as group two, you know that $\bar{x} = 35{,}328$; $\sigma_1 = 750$; $n_1 = 100$; $\bar{y} = 30{,}250$; $\sigma_2 = 500$; and $n_2 = 100$ Your Z^* value is 1.96 (from Table 11-1). Using the formula for a confidence interval for the difference of two population means, the 95 percent confidence interval is $(\bar{x}-\bar{y}) \pm Z^* * \sqrt{\frac{\sigma_1^2}{n_1}+\frac{\sigma_2^2}{n_2}} =$ $(35{,}328 - 30{,}250) \pm 1.96^* * \sqrt{\frac{750^2}{100}+\frac{500^2}{100}} =$ $5{,}078 \pm 1.96^* * 90.14 = 5{,}078 \pm 176.67$ miles. *Interpretation:* You're 95 percent confident that the difference in average mileage between all cars on Lots 1 and 2 is $5{,}078 \pm 176.67$ miles (and that cars on Lot 1 have higher miles than cars on Lot 2).

13 Based on a sample of 100 participants, the average weight loss the first month under a new (competing) weight-loss plan is 11.4 pounds with a population standard deviation of 5.1 pounds. The average weight loss for the first month for 100 people on the old (standard) weight-loss plan is 12.8 pounds, with population standard deviation of 4.8 pounds.

a. Find a 90 percent confidence interval for the difference in weight loss for the two plans (old minus new).

b. What's the margin of error for your calculated confidence interval?

14 The average miles per gallon for a sample of 40 cars of model SX last year was 32.1, with a population standard deviation of 3.8. A sample of 40 cars from this year's model SX has an average of 35.2 mpg, with a population standard deviation of 5.4.

a. Find a 99 percent confidence interval for the difference in average mpg for this car brand (this year's model minus last year's).

b. Find a 99 percent confidence interval for the difference in average mpg for last year's model minus this year's. What does the negative difference mean?

15 You want to compare the average number of tines on the antlers of male deer in two nearby metro parks. A sample of 30 deer from the first park shows an average of 5 tines with a population standard deviation of 3. A sample of 35 deer from the second park shows an average of 6 tines with a population standard deviation of 3.2.

a. Find a 95 percent confidence interval for the difference in average number of tines for all male deer in the two metro parks (second park minus first park).

b. Do the parks' deer populations differ in average size of deer antlers?

Suppose that a group of 100 people is on one weight-loss plan for one month and then switches to another weight-loss plan for one month. Suppose that you want to compare the two plans, and you decide to compare the average weight loss for the first month to the second month. Explain why you can't use this data to make a 95 percent confidence interval for the average difference using the formulas from this section.

Computing a Confidence Interval for the Difference of Two Proportions

You use a confidence interval for the difference in two population proportions under three conditions:

- The variable you deal with is categorical, such as gender, political affiliation, opinion (agree, disagree, no opinion), or marital status.
- You want to estimate the difference between the two population proportions.
- The sample sizes are large enough.

Because you want to compare two separate (independent) populations, you need two separate (independent) samples — one from each population. The formula for a $(1-\alpha)$ percent confidence interval for the difference of two population proportions is

$$(\hat{p}_1 - \hat{p}_2) \pm Z^* * \sqrt{\frac{\hat{p}_1(1-\hat{p}_1)}{n_1} + \frac{\hat{p}_2(1-\hat{p}_2)}{n_2}}.$$

Besides the usual Z^* value, you need two pieces of information from each sample to calculate this confidence interval: the sample proportion, $\hat{p}$, and the sample size (n). For the first sample, the sample proportion and the sample size are denoted $\hat{p}_1$ and n_1, respectively. For the second sample, the sample proportion and the sample size are denoted $\hat{p}_2$ and n_2, respectively.

The sample sizes are large enough if you meet two sets of conditions: (1) $n_1 * \hat{p}_1$ and $n_1 * (1-\hat{p}_1)$ are at least 10; and (2) $n_2 * \hat{p}_2$ and $n_2 * (1-\hat{p}_2)$ are at least 10. You meet the conditions in the vast majority of survey situations (where you mainly use this confidence interval). To calculate the confidence interval here:

1. **Take the sample proportion for the first group minus the sample proportion for the second group to get your sample statistic (for the difference).**
2. **Add and subtract the margin of error to get your confidence interval.**
3. **Interpret your results.**

See the following for an example of finding a confidence interval for the difference in two population proportions.

EXAMPLE

Q. A random sample of 500 male U.S. college students finds that 35 percent watch the Super Bowl every year, and a random sample of 500 female U.S. college students finds that 21 percent watch every year. Find a 95 percent confidence interval for the difference in the proportion of Super Bowl watchers for male versus female U.S. college students.

A. For the males, the sample proportion is 0.35, and $n = 500$. For females, the sample proportion is 0.21, and $n = 500$. You know $Z^* = 1.96$ (from Table 11-1). The 95 percent confidence interval is $(\hat{p}_1 - \hat{p}_2) \pm Z^* * \sqrt{\frac{\hat{p}_1(1-\hat{p}_1)}{n_1} + \frac{\hat{p}_2(1-\hat{p}_2)}{n_2}} = (0.35 - 0.21) \pm 1.96 *$ $\sqrt{\frac{0.35(1-0.35)}{500} + \frac{0.21(1-0.21)}{500}} = 0.14 \pm 1.96 * \sqrt{\frac{0.35(1-0.35)}{500} + \frac{0.21(1-0.21)}{500}} =$ $0.14 \pm 1.96 * \sqrt{.000455 + .00033} = 0.14 \pm 1.96 * 0.028 = 0.14 \pm 0.055 = (0.085,\ 0.195)$.

Interpretation: You're 95 percent confident that the difference in the proportion of Super Bowl watchers for male versus female U.S. college students is between 0.085 and 0.195. In other words, between 8.5 percent and 19.5 percent more males watch the Super Bowl than females.

17 A random sample of 1,117 domestic students at a U.S. university finds that 915 go home at least once each term, compared to 212 from a random sample of 1,200 international students from the same university. Find a 98 percent confidence interval for the difference in the proportion of students who go home at least once each term.

18 A poll of 1,000 smokers shows that 16 percent approve of a smoking ban in bars and restaurants; 84 percent from a sample of 500 nonsmokers approve of the ban.

a. What's the margin of error for a 95 percent confidence interval for the difference in proportion for all smokers versus all nonsmokers?

b. In this case, the sample percents sum to 1; does this always happen?

19 Suppose that 72 percent of a sample of 1,000 traditional college students drive a used car as opposed to a new car or no car at all. A sample of 1,000 nontraditional students shows 71 percent drive used cars.

a. Find an 80 percent confidence interval for the difference in the percentages of students who drive used cars for these two populations.

b. The lower bound of this interval is negative, and the upper bound is positive. How do you interpret these results?

20 Suppose that a class of 100 students contains 10 percent Independents and 90 percent party affiliates. If Bob samples half of the students in each group and compares the percentage approving of the president, can he use the formula for finding a confidence interval for the difference of two proportions? Why or why not?

Answers to Problems in Calculating Confidence Intervals

1. The margin of error is the part you add and subtract to get the upper and lower boundaries of the confidence interval.

   a. The margin of error here is plus or minus 35 minutes.

   b. The lower boundary is $110-35=75$ minutes; the upper boundary is $110+35=145$ minutes.

2. Increasing confidence increases the width of the confidence interval.

   a. The 99.7 percent confidence interval is wider, because (all else remaining the same) the Z^* value increases from 1.96 to 2.96 (see Table 11-1), which increases the margin of error.

   b. A wider interval means you aren't as precise at estimating your population parameter — not a good thing.

3. No. You're 100 percent confident that the sample statistic is in the confidence interval, because you find the confidence interval by using the sample statistic plus or minus the margin of error.

4. No. It means that 95 percent of all intervals obtained by different samples contain the population parameter. One particular 95 percent confidence interval either contains it or it doesn't.

TIP

As a professor, I'm 95 percent confident that your instructor will ask you what a 95 percent confidence interval means; all statistics professors love to use this doozy on exams, so be ready!

5. The question asks you to calculate two different confidence intervals here.

   a. The 95 percent confidence interval for the average number of tines on deer in the park is 5, plus or minus $Z^* * \frac{\sigma}{\sqrt{n}} = 1.96 * \frac{3}{\sqrt{30}} = 1.07$. *Interpretation:* You're 95 percent confident that the average number of tines on the deer in this park is 5 ± 1.07. That is, between 3.93 and 6.07.

   b. The 98 percent confidence interval is 5 plus or minus $Z^* * \frac{\sigma}{\sqrt{n}} = 2.33 * \frac{3}{\sqrt{30}} = 1.28$. *Interpretation:* You're 98 percent confident that the average number of tines on the deer in this park is 5 ± 1.28. That is, between 3.72 and 6.28.

6. Try to interpret confidence intervals in a way that a layperson can understand.

   a. According to your sample, the average weight loss for everyone in the program in month one is 11.4 pounds, plus or minus 0.51 (or (10.89, 11.91) pounds). Your process of sampling will yield the correct answer 95 percent of the time.

   b. The margin of error is plus or minus 0.51 pound.

7. Margin of error maintains the same units as the original data. Be aware that sample size strongly affects the margin of error.

   a. Miles per gallon.

   b. Increase the sample size.

8. Yes, by also increasing the sample size. That will offset the larger Z^* value that goes with the 95 percent confidence level.

REMEMBER

The margin of error increases when the confidence level or standard deviation goes up and decreases when the sample size goes up (refer to Chapter 10).

9. The needed pieces of this problem are $n = 1.117$, $\hat{p} = 729 \div 1{,}117 = 0.65$, and $Z^* = 2.33$ (from Table 11-1) for a 98 percent confidence level. The confidence interval is $\hat{p} \pm Z^* * \sqrt{\frac{\hat{p}(1-\hat{p})}{n}} = 0.65 \pm 2.33 * \sqrt{\frac{0.65(1-0.65)}{1{,}117}} = 0.65 \pm 2.33 * \sqrt{0.0002} = 0.65 \pm 0.03 = (0.62,\ 0.68)$.

   Interpretation: You're 98 percent confident that the proportion of all U.S. college students who go home at least once each term is between 0.62 and 0.68.

WARNING

Be careful with what you use in the formula for $\hat{p}$. Don't use the total number in the category of interest (in this example, 729). You need to divide the total number by the total sample size to get an actual proportion (in this case, $729 \div 1{,}170 = 0.65$) — a number between zero and one — or you get incorrect calculations. Also, you can't use the percentage in the formula; use the proportion (decimal version of the percent). In this example, you use 0.65, not 65.

10. Here you have $n = 2{,}500$, $\hat{p} = .50$, and $Z^* = 1.96$ (from Table 11-1), so the margin of error is plus or minus $Z^* * \sqrt{\frac{\hat{p}(1-\hat{p})}{n}} = \pm 1.96 * \sqrt{\frac{0.50(1-0.50)}{2{,}500}} = \pm 1.96 * 0.01 = \pm 0.020$.

11. Confidence levels and sample sizes affect confidence intervals.

    a. Here you have $n = 1{,}000$, $\hat{p} = 0.73$, and $Z^* = 1.28$ (from Table 11-1), so the 80 percent confidence interval is $\hat{p} \pm Z^* * \sqrt{\frac{\hat{p}(1-\hat{p})}{n}} = 0.73 \pm 1.28 * \sqrt{\frac{0.73(1-0.73)}{1{,}000}} = 0.73 \pm 1.28 * 0.014$ and $0.73 \pm 0.0179 = (0.71,\ 0.75)$, which is 71 percent to 75 percent. *Interpretation:* You're 80 percent confident that the percentage of all U.S. college students who drive a used car is between 71 percent and 75 percent. (***Note:*** I used the decimal versions for all calculations and then changed to percentages at the very end.)

    b. Quadruple the sample size to 4,000. You don't need calculations (refer to Chapter 10).

12. No, because the group uses a biased sample, so the margin of error is meaningless (not legitimate).

TIP

Instructors are big on the idea "Garbage in equals garbage out." It applies to confidence intervals (and basically every other statistical calculation). If you put bad data into a formula, the formula still cranks out an answer, but the answer may be garbage. You have to evaluate results by the stated margin of error and by the manner of data collection.

13 Keeping track of which group is which, and staying consistent throughout your calculations, will help you a great deal on these difference of means problems.

a. Using the notation and letting the old plan be group one, you have $\bar{x} = 12.8$, $\sigma_1 = 4.8$, $n_1 = 100$, $\bar{y} = 11.4$, $\sigma_2 = 5.1$, $n_2 = 100$, and $Z^* = 1.64$. Therefore, the 90 percent confidence interval for the difference in average weight loss (old plan minus competing plan) is

$$(\bar{x} - \bar{y}) \pm Z^* * \sqrt{\frac{\sigma_1^2}{n_1} + \frac{\sigma_2^2}{n_2}} = (12.8 - 11.4) \pm 1.64 * \sqrt{\frac{4.8^2}{100} + \frac{5.1^2}{100}} = 1.4 \pm 1.64 * 0.700 = 1.4 \pm 1.1.$$

Interpretation: You're 90 percent confident that the difference in average weight loss on the two plans is 1.4 ± 1.1 pounds. That is, those on the standard (old) plan lose an average of between 0.3 and 2.5 more pounds than those on the competing (new) plan.

Note: Depending on which group you choose to make group one, your results are completely opposite (in terms of sign) if you switch the groups, which is fine as long as you know which group is which. You may want to always choose the group with the highest mean as group one so the difference between the sample means is positive rather than negative.

b. The margin of error is the part after the ± sign, which in this case is 1.1 pounds.

14 Switching the order of groups switches the sign on the upper and lower bounds of your confidence interval.

a. Using the notation, you have $\bar{x} = 35.2$, $\sigma_1 = 5.4$, $n_1 = 40$, $\bar{y} = 32.1$, $\sigma_2 = 3.8$, $n_2 = 40$, and $Z^* = 2.58$, so the 99 percent confidence interval for this year minus last year is $(\bar{x} - \bar{y}) \pm Z^* * \sqrt{\frac{\sigma_1^2}{n_1} + \frac{\sigma_2^2}{n_2}} =$

$$(35.2 - 32.1) \pm 2.58 * \sqrt{\frac{5.4^2}{40} + \frac{3.8^2}{40}} = 3.10 \pm 2.58 * 1.044 = 3.10 \pm 2.69 \text{ miles per gallon.}$$

Interpretation: You're 99 percent confident that the difference in average mpg for this year's brand versus last year's is 3.10 ± 2.69. That is, this year's model gets on average between 0.41 and 5.79 more miles per gallon than last year's model.

b. In this case, you find last year minus this year, so the 99 percent confidence interval switches the sample means (all else remains the same), which gives you -3.10 ± 2.69 miles per gallon. That is, (–5.79,–0.41). You still get a correct answer; it just may be harder to interpret. Because the differences are negative in this confidence interval, that means the first group (last year's model) had fewer miles per gallon than the second group (this year's model).

TIP

If the difference between two numbers is negative, the second number is larger than the first. Use this added piece of information in your interpretation of the results. Not only do you know that the groups differ, but you also know which one has the larger value and which one has the smaller value.

15 Just because two sample means are different doesn't mean you should expect their population means to be different. It all depends on what your definition of "different" is.

REMEMBER

Sample results are always going to vary from sample to sample, so their differing is no big deal. The question is, are they different enough that, even taking that variability into account, you can be confident that one will be higher than the other in the population? That's what being statistically significant really means.

a. Use the second park as group one to keep the numbers positive. Using the notation, you have $\bar{x} = 6$, $\sigma_1 = 3.2$, $n_1 = 35$, $\bar{y} = 5$, $\sigma_2 = 3$, $n_2 = 30$, and $Z^* = 1.96$, so the 95 percent confidence interval is $(\bar{x} - \bar{y}) \pm Z^* * \sqrt{\frac{\sigma_1^2}{n_1} + \frac{\sigma_2^2}{n_2}} = (6-5) \pm 1.96 * \sqrt{\frac{3.2^2}{35} + \frac{3.0^2}{30}} = 1 \pm 1.96 * 0.77 = 1 \pm 1.51$, or $(-0.51, 2.51)$ tines per deer. *Interpretation:* You're 95 percent confident that the difference in average number of tines for all male deer in the two metro parks is between –0.51 and 2.51. That is, you can't say the deer in one park have more tines on average than the other because the difference could go either way, depending on the sample that's taken.

b. No, because the lower boundary is positive, the upper boundary is negative, and zero is in the interval. Therefore, as samples vary, the estimates for the differences in the two populations are "too close to call," so you conclude no significant difference in the two population means.

16 You need two independent samples to use the confidence interval for the difference of two populations, and the two samples aren't independent. The same people participate in both samples.

17 Using the notation, you have $\hat{p}_1 = \frac{915}{1{,}117} = 0.82$, $n_1 = 1{,}117$, $\hat{p}_2 = \frac{212}{1{,}200} = 0.18$, $n_2 = 1{,}200$, and $Z^* = 2.33$, so the 98 percent confidence interval is $(\hat{p}_1 - \hat{p}_2) \pm Z^* * \sqrt{\frac{\hat{p}_1(1-\hat{p}_1)}{n_1} + \frac{\hat{p}_2(1-\hat{p}_2)}{n_2}} =$ $(0.82 - 0.18) \pm 2.33 * \sqrt{\frac{0.82(1-0.82)}{1{,}117} + \frac{0.18(1-0.18)}{1{,}200}} = 0.64 \pm 2.33\sqrt{0.00013 + 0.000123} =$ $0.64 \pm 0.04 = (0.60, 0.68)$. *Interpretation:* You're 98 percent confident that the difference in the proportion of all college students who go home at least once each term (domestic versus international students) is between 0.60 and 0.68. And because the differences are positive, this result says that the domestic students go home that much more often (60 to 68 percent more) than the international students.

18 This is a standard question that asks you to find a confidence interval for the difference in two population proportions.

a. Assign group one to the nonsmokers. Using the notation, you have $\hat{p}_1 = 0.84$, $n_1 = 500$, $\hat{p}_2 = 0.16$, $n_2 = 1{,}000$, and $Z^* = 1.96$, so the 95 percent confidence interval for the difference in population proportions is $(\hat{p}_1 - \hat{p}_2) \pm Z^* * \sqrt{\frac{\hat{p}_1(1-\hat{p}_1)}{n_1} + \frac{\hat{p}_2(1-\hat{p}_2)}{n_2}} =$

$(0.84 - 0.16) \pm 1.96 * \sqrt{\frac{0.84(1-0.84)}{500} + \frac{0.16(1-0.16)}{1{,}000}} = 0.68 \pm 1.96\sqrt{0.00027 + 0.00013} =$ $0.68 \pm 1.96 * 0.02 = 0.68 \pm 0.04 = (0.64, 0.72)$.

Interpretation: You're 95 percent confident that the difference in the proportion of smokers versus nonsmokers who approve the smoking ban is 0.68 ± 0.04. In percentage terms, that means 68 percent more nonsmokers approve of the ban than smokers, plus or minus 4 percent.

b. No. The proportion from Sample 1 that falls under the desired category of interest has nothing to do with the proportion from Sample 2 that does. Remember, the samples are independent.

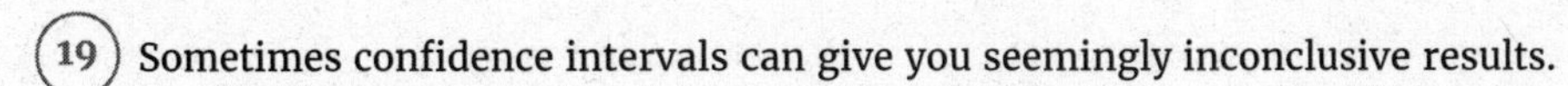

19 Sometimes confidence intervals can give you seemingly inconclusive results.

a. Assign the traditional students to group one. Using the notation, you have $\hat{p}_1 = 0.72$, $n_1 = 1{,}000$, $\hat{p}_2 = 0.71$, $n_2 = 1{,}000$, and $Z^* = 1.28$, so the 80 percent confidence interval for the difference in population proportions is $(\hat{p}_1 - \hat{p}_2) \pm Z^* * \sqrt{\frac{\hat{p}_1(1-\hat{p}_1)}{n_1} + \frac{\hat{p}_2(1-\hat{p}_2)}{n_2}} =$

$$(0.72 - 0.71) \pm 1.28 * \sqrt{\frac{0.72(1-0.72)}{1{,}000} + \frac{0.71(1-0.71)}{1{,}000}} = 0.01 \pm 1.28\sqrt{0.00020 + 0.00021} =$$

$0.01 \pm 0.03 = (-0.02,\ 0.04)$.

Interpretation: You're 95 percent confident that the difference in the percentages of students who drive used cars for traditional students versus nontraditional students is between −0.02 and 0.04. This result means you can't say one group drives used cars more often than the other group.

b. Because zero is included in this interval, the difference is "too close to call," so you conclude no significant difference between the two populations.

20 No. Because the total population is 100, and 10 percent are Independent, the class produces only 10 Independents. Because Bob samples half of the 10, his sample size is only 5 Independents. Because $n * \hat{p} = 5 * 0.10 = 0.5$ isn't at least 10 for the Independent group, the sample size condition for this confidence interval formula isn't met. Bob should take a larger sample of Independents for his study.

» Avoiding common interpretation mistakes

» Knowing what confidence intervals don't measure

Chapter 12
Deciphering Your Confidence Interval

Confidence intervals can be straightforward, black-and-white calculations (with practice), yet their real meaning can still remain a mystery. With the guidance in this chapter, I help you avoid making the mistakes that drive instructors (and, subsequently, you) crazy. You can become one of the enlightened who "get it" and thereby get more points on exams. Because in the end, it's all about points, isn't it?

Interpreting Confidence Intervals the Right Way

Because I've been a statistics professor for many years, I know how disappointed instructors can get when students don't completely understand what the results of a confidence interval really mean. (Part of it is our own fault because we have a hard time explaining it clearly ourselves.) Instructors want to harp and wax philosophic on this issue to no end, and we make it a personal challenge to help you understand it. In other words, confidence intervals show up on every quiz or exam that instructors can possibly include them on. The practice problems in this chapter put you through every scenario I can think of to help you avoid misinterpreting a confidence interval. I also address some big picture ideas regarding confidence intervals and how to view them.

Here's the scoop on what a confidence interval really means: Suppose that you just found a 95 percent confidence interval for the population mean. To interpret it correctly, you say, "We are 95 percent confident that the population mean is in this interval." That does not mean, however, that there is a 95 percent chance that the population mean is in your interval. After the interval has been calculated, it's either right or wrong, and you don't put a probability on it. The 95 percent is how confident you are in the process that led you to the interval you got; in other words, your sampling process. It means that 95 percent of the time, you will get random samples that do represent the situation and result in a correct interval (that contains the population parameter). Five percent of the time you won't, just by chance. Read this paragraph over a couple of times and really think hard about it before moving on to the problems. You are at the crux of the issue right here, and it may take some time to really grab hold of the idea. Take your time.

See the following for an example of interpreting a confidence interval.

Q. Suppose that survey results say that 69 percent of adult females from a sample taken in the United States registered to vote. The margin of error for the survey that produced these results was plus or minus 3 percent (based on 95 percent confidence). Interpret the results of this confidence interval.

A. Based on your sample, you're 95 percent confident that the percentage of all females in the United States who were registered to vote was between 66 percent and 72 percent. (This is because $69 - 3 = 66$, and $69 + 3 = 72$.) The results also mean that if you repeat this survey with many different samples of the same size, 95 percent of the samples would result in intervals that contain the true percentage of females who registered to vote. Better hope that yours is one of those intervals!

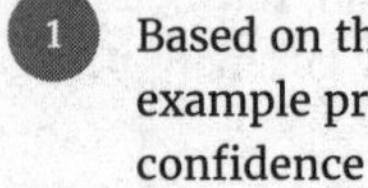

1. Based on the results of the survey from the example problem, can you say with 95 percent confidence that your confidence interval contains the true percentage of all adult females registered to vote?

2. Based on the results of the survey from the example problem, can you say that 69 percent of all adult women in the United States registered to vote?

3 Based on the results of the survey from the example problem, can you say that 69 percent of all adult female respondents to this survey registered to vote?

4 Based on the results of the survey from the example problem, can you say that it's probably true that 69 percent of all female adults in the United States registered to vote?

5 Based on the results of the survey from the example problem, can you say that the percentage of all adult females in the United States who registered to vote is between 66 and 72 percent?

6 Based on the results of the survey from the example problem, can you say with 95 percent confidence that the interval 69 percent plus or minus 3 percent contains the percentage of registered voters for adult females in the sample?

7 Based on the results of the survey from the example problem, can you say that the true proportion of adult female registered voters in the United States has a 95 percent chance of falling somewhere between 66 and 72 percent?

8 Based on the results of the survey from the example problem, can you say that if you repeat this same sampling process over and over with the same sample size, you would get an incorrect confidence interval only 5 percent of the time (that is, it would not contain the true proportion of registered voters for all adult females in the United States)?

9 Is a wide confidence interval a good thing?

10 Will confidence intervals with a high confidence always be wide?

11 Does a larger confidence level lower the chance for bias in the results?

12 Bob thinks that to cut his margin of error in half, he needs to sample twice as many people. Is Bob right?

Evaluating Confidence Interval Results: What the Formulas Don't Tell You

When data comes from well-designed surveys and experiments, and is based on large random samples, you can feel good about the quality of the information. When the margin of error of any confidence interval is small, you assume that the confidence interval provides an accurate and credible estimate of the parameter. This isn't always the case, however. Why not? Because not all data come from well-designed surveys and experiments and are based on large random samples.

REMEMBER

When it comes to margin of error, less may be more. The formulas don't realize it when the numbers plugged into them are based on biased data, so you have to spot those situations and disregard the seemingly precise results.

See the following for an example of when a reported margin of error is meaningless.

EXAMPLE

Q. Suppose that a survey on a popular Internet website receives responses from 50,000 people. The reported margin of error for this survey, according to the formula, is about 0.0045, or 0.45 percent, which is tiny. Is this margin of error correct? Explain. Assume that the calculations are correct.

A. According to the formulas, the mathematics may be correct, but the results are based on biased data and are therefore bogus. Basically, all Internet polls are bogus except those that actually go out and select their participants at random from the population. And that is impossible to do with a general population because many folks don't use computers or go online.

13 Does margin of error measure bias?

14 Suppose that a margin of error isn't reported. Should you automatically assume that it has a small value and move on?

Answers to Problems in Confidence Intervals

1. Yes, you can say that you are 95 percent confident that the population parameter is in your confidence interval.

TIP

Instructors like to make a big deal out of this interpretation issue. Avoid the trap of saying that a 95 percent confidence interval means that the parameter has a 95 percent chance of being in your interval. After you create the interval, the parameter either falls in or it doesn't. The 95 percent confidence is in your sampling process, before the fact. After the interval is done, the parameter is either in there or it isn't.

2. No, the only thing you can be sure of is that 69 percent of the *sample* registered to vote. You can't put a one-number guess on where the population percentage lies; you need an interval of possible values (hence, a confidence interval).

WARNING

You should never give a one-number estimate as to what the value of a population parameter is. You always need an interval, which includes the one-point estimate plus or minus a margin of error.

3. Yes, because this statement is about the sample, which you know everything about. However, the statement doesn't mean much to you, because ultimately you want to find out about the target population (all females in the United States), not just those in the sample.

4. No, because the figure may be 69 percent, or it may be another percentage. You can hope that the population percentage is close to 69 percent, but you have no guarantee of that.

5. This problem makes you focus on the lower and upper limits of the confidence interval, which are correct, because 69 percent plus or minus 3 percent gives you 66 percent and 72 percent for your limits. However, you can't say for sure that the actual population percentage falls in between — that you don't know.

6. Well, you can be 100 percent sure that the percentage falls in between those values. Why? Because this statement concerns the percentage in the sample, you don't need an interval. You know the percentage is exactly 69. (Again, this kind of statement isn't relevant, because the purpose of a confidence interval is to give you some idea of where the population parameter may be, using the sample statistic.)

7. No, because this statement again puts a probability on whether or not the parameter is in the interval. The probability should pertain to the intervals, not to the parameters. In 95 percent of the intervals, the parameter is included, because 95 percent of the random samples represent the true population and 5 percent don't, just by chance.

8. Yes, you can say that. Just hope your interval isn't one of those 5 percent.

9. No. The goal is to have a narrow confidence interval with a high level of confidence, because that means less chance for error while you (hopefully) zoom in close on the true value.

10. Not necessarily. If the sample size is large, it decreases the margin of error and can offset the larger confidence level that requires a larger Z^* or t-value.

11. No. Bias isn't measured by any part of a confidence interval.

12. No, Bob is wrong. If you look at the formula for margin of error for a proportion $Z^* * \left(\sqrt{\frac{\hat{p}(1-\hat{p})}{n}}\right)$, you see a square root of n in the denominator. That means that 4 times the sample size decreases the margin of error by a factor of 2 (the square root of 4). So, to cut the margin of error in half, you need to quadruple the sample size (multiply it by 4).

13. No. Bias is the amount by which the statistic in the front part of the confidence interval can be systematically off (on the high end or the low end). Bias is nearly impossible to measure, and the equations for margin of error certainly don't include it.

WARNING

Margin of error doesn't measure bias. Biased results cause inaccurate statistics, and no matter how small the added and subtracted margin of error seems to be, the whole interval is inaccurate due to the bias. The best policy is to avoid bias and discount results that seem precise but are based on biased data.

14. No, you can't assume that it's a small value. If you have the sample size and the percentage from the sample, you can figure out the size for yourself, but for quantitative data, you need both the sample mean and standard deviation.

REMEMBER

Don't assume that "no margin of error means a good (small) margin of error." There may be a reason the margin isn't reported, and that reason may not be good.

» Setting up your null and alternative hypotheses

» Calculating test statistics

» Determining the critical values

» Drawing conclusions and making interpretations

Chapter 13
Testing Hypotheses

A *hypothesis test* is a statistical procedure designed to test a claim about a population. In other words, someone proposes a certain parameter, and you want to test the claim by using data. Testing a hypothesis is different from a confidence interval situation, where you have no idea what the parameter is and you use your data to estimate it.

In this chapter, you review the basic ideas and steps of conducting a hypothesis test, and you practice setting up and carrying out the most common hypothesis tests: the tests for a population mean, a population proportion, two population means, and two population proportions. You also practice working with the t-distribution, which you use in situations where the sample sizes are too small to use the Z-distribution, or if the population standard deviation, σ, is unknown (see Chapters 7 and 8 for more on the Z- and t-distributions).

The most common hypothesis test is the test for one population mean. Someone is claiming that the population mean is one value, and you are testing that claim because you believe it to be false. In this first section, you go through each step of a hypothesis test using the situation where you are testing one population mean. You also assume that the population standard deviation, σ, is known and that you have either a normal distribution for your population or a large enough sample size (sample $n > 30$).

Walking Through a Hypothesis Test

Every hypothesis test contains two hypotheses. The first hypothesis is called the *null hypothesis*, denoted *Ho* (pronounced "H naught"). The null hypothesis states that the population parameter is equal to the claimed value. For example, if you claim that the average test score for a population of students is 75, you have Ho: $\mu = 75$. In general, when you are conducting a hypothesis test about the population mean, you use μ_o to denote the claimed value of μ in the null hypothesis (for example, 75).

Along with every null hypothesis is an alternative hypothesis, denoted *Ha* (or H_1). If Ho turns out to be false, you conclude that the alternative hypothesis is true. Three possibilities exist for the second or alternative hypothesis, denoted Ha:

- The population parameter is *not equal to* the claimed value, written as Ha: $\mu \neq \mu_o$.
- The population parameter is *greater than* the claimed value, denoted Ha: $\mu > \mu_o$.
- The population parameter is *less than* the claimed value, denoted Ha: $\mu < \mu_o$.

Which alternative hypothesis you choose when setting up your hypothesis test depends on what you want to conclude, should you have enough evidence to refute the claim (Ho). If someone claims that the average test score is higher than 75, you have Ha: $\mu > 75$. If someone thinks that the percentage of students who own cellphones is less than 65 percent, you have Ha: $p < 0.65$.

TIP

The null hypothesis can be written either as Ho or as H_o, depending on who's writing it, so don't be confused if you see it one way in your textbook and another way in this workbook. I write it as Ho in this workbook; ditto with the alternative hypothesis, which I denote as Ha.

After you set up the hypotheses, the next step is to collect the data and calculate your sample statistic (the sample mean, $\bar{x}$). Convert your statistic to a standard score so you can interpret it on a Z-table (see the Appendix). To convert your sample statistic to a test statistic:

1. **Take your statistic minus the number given by Ho, $\bar{x} - \mu_o$.**
2. **Divide the result by the standard error of the statistic.**

   Assuming you have a normal distribution and the population standard deviation is known, the standard error for the sample mean is $\frac{\sigma}{\sqrt{n}}$ (see Chapter 10).

   The test statistic is $Z = \frac{\bar{x} - \mu_o}{\sigma/\sqrt{n}}$.

After you calculate your test statistic, the hard part is over. Now all you have to do is make your conclusion by seeing where the test statistic falls on the Z-distribution. You can do this in one of two ways: using critical values or *p*-values. This chapter deals with the critical value method; the *p*-value method is in Chapter 14.

Under the critical value method, before you collect your data, you set one (or two) cutoff point(s) on the Z-distribution so that if your test statistic falls beyond the cutoff point(s), you reject Ho; otherwise, you fail to reject Ho. The cutoff points are called *critical values.* (A critical value is much like the goal line in football; the place you have to reach to "score" a significant result and reject Ho.)

The critical value(s) is (are) determined by the significance level (or alpha level) of the test, denoted by α. Alpha levels differ for each situation, but most researchers are happy with an alpha level of 0.05, much in the same way they're happy with a 95 percent confidence level for a confidence interval. (Notice that $1-\alpha$ equals the confidence level of a confidence interval.)

On exams, oftentimes your instructor will tell you which alpha level she wants you to use for a particular problem. If she doesn't state it, 0.05 is the most common one to use, and I would consider it a safe bet.

Table 13-1 shows some critical values for one-sided and two-sided hypothesis tests that use the Z-distribution. My computer software generated the values, so they're more precise than what you find in the Z-table (refer to the Appendix). Many more alpha levels are possible, but this table gives you a good start. Note the not-equal-to alternative has $1-\alpha$ of the probability lying between two values, not less than one value (as in the > alternative) and not all greater than one value (as in the < alternative).

Table 13-1 Critical Values for Hypothesis Tests Using the *Z*-Distribution

Alpha Level (α)	Alternative Hypothesis	Critical Value(s)
0.01	>	+2.33
0.01	<	–2.33
0.01	≠	–2.58 and +2.58
0.05	>	+1.64
0.05	<	–1.64
0.05	≠	–1.96 and +1.96
0.10	<	–1.28
0.10	>	+1.28
0.10	≠	–1.64 and +1.64

After you set up the critical value(s), if the test statistic falls beyond the critical value(s), your conclusion is "reject Ho at level α." This means that the test statistic falls into the *rejection region*. If the test statistic doesn't go beyond the critical value(s), you conclude "fail to reject Ho at level α." This means that the test statistic falls into the *nonrejection region.*

I show the *p*-value method for making your conclusions in a hypothesis test in Chapter 14. Statisticians often prefer this method, because it allows you to report how strong your evidence actually is instead of simply stating whether you reject Ho at a certain alpha level.

See the following for an example of setting up and conducting a hypothesis test for a population mean.

EXAMPLE

Q. Suppose that you test Ho: $\mu = 7$ versus Ha: $\mu < 7$, and your sample mean is 6.5 with a sample of 10, and the population standard deviation is 0.5. Assume that your data come from a normal distribution.

a. What's the value of μ_o?

b. What type of test is this: a right-tailed, left-tailed, or two-tailed test?

c. What's the critical value if you use $\alpha = 0.01$? (Assume a normal distribution.)

d. What's your test statistic?

e. What's your conclusion?

A. You use this test when you're trying to estimate a population mean.

a. The value of μ_o is 7.

b. A left-tailed test, because Ha has a "<" sign in it.

c. The critical value is –2.33, using Table 13-1.

d. The test statistic is $Z = \frac{\bar{x} - \mu_o}{\sigma / \sqrt{n}} = \frac{6.5 - 7}{0.5 / \sqrt{10}} = -3.16$.

e. The conclusion is "reject Ho" because the test statistic is beyond the critical value of –2.33 and hence falls in the "rejection" region.

1 Explain what's wrong with the following hypotheses: Ho: $\bar{x} = 7$ versus Ha: $\bar{x} \neq 7$.

2 Suppose that a pizza place claims its average pizza delivery time is 30 minutes, but you believe it takes longer than that. Your sample of 10 pizzas has an average delivery time of 40 minutes. Assume that the population standard deviation is 15 minutes and the times have a normal distribution. Use $\alpha = 0.05$.

a. What are your null and alternative hypotheses?

b. What is the critical value?

c. What is the test statistic?

d. What is the conclusion?

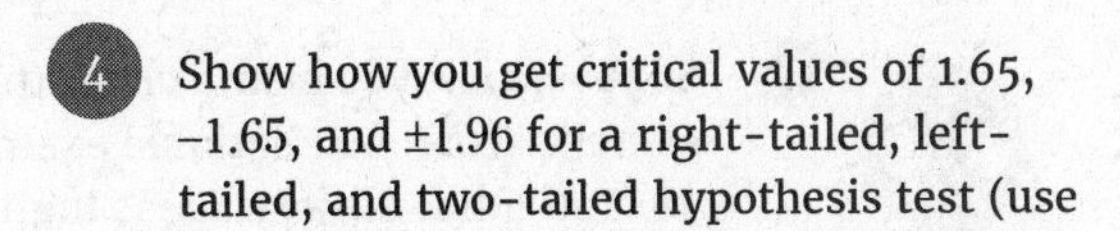

3 Suppose that a sports reporter claims the average football game lasts 3 hours, and you believe it's more than that. Your random sample of 35 games has an average time of 3.25 hours. Assume that the population standard deviation is 1 hour. Use $\alpha = 0.05$. What do you conclude?

4 Show how you get critical values of 1.65, -1.65, and ± 1.96 for a right-tailed, left-tailed, and two-tailed hypothesis test (use $\alpha = 0.05$ and assume a large sample size).

Testing a Hypothesis about a Population Mean

It is not always the case that σ is known or the population has a normal distribution or n is large enough to the use the central limit theorem (CLT). The formula for the test statistic for one population mean in those cases is $t = \frac{\bar{x} - \mu_o}{\frac{s}{\sqrt{n}}}$. Here, s is the sample standard deviation. To calculate it, you

1. **Calculate the sample mean, $\bar{x}$, and use the given population standard deviation, s. Let n represent the sample size.**
2. **Calculate the standard error, $\frac{s}{\sqrt{n}}$. Save your answer.**
3. **Find the sample mean minus μ_o.**
4. **Divide your result from Step 3 by the standard error you find in Step 2.**

Compare your test statistic to the critical value(s) from the t-distribution with $n-1$ degrees of freedom, as described in Chapter 8. Use the t-table (see the Appendix). If σ is unknown, or the sample size is small, use the sample standard deviation, s, instead of σ and use the t-distribution with $n-1$ degrees of freedom. If your test statistic is beyond the critical value(s), reject Ho; otherwise, fail to reject Ho.

REMEMBER

To use Z in a hypothesis test, you need either a normal distribution to start with or n to be at least 30 and use the CLT. If the sample size is < 30, you can use the t-distribution. In this case, the test is called a *t-test*.

See the following for an example of setting up a t-test for the mean.

EXAMPLE

Q. Suppose that you hear a claim that the average score on a national exam is 78. You think the average is higher than that, and your sample of 100 students produces an average of 81.

a. Set up your null and alternative hypotheses.

b. Find the critical value when $\alpha = 0.01$.

A. Setting up the hypotheses correctly is critical to your success with hypothesis tests.

a. In this case, you have Ho: $\mu = 78$ versus Ha: $\mu > 78$. ***Note:*** 81 is a sample statistic and doesn't belong in Ho or Ha. You want to show that the mean is higher than the claim, so the alternative has a ">" sign. You're conducting a right-tailed test.

b. The critical value is $t = 2.33$. Here's how you get it: We use Table 13-1 here because $n = 100$ is so large that the t table and Z-table would give the same values.

5 Conduct the hypothesis test Ho: $\mu = 7$ versus $\mu > 7$, where $\bar{x} = 7.5$, $s = 2$, and $n = 30$. Use $\alpha = 0.01$.

6 Conduct the hypothesis test Ho: $\mu = 75$ versus Ho: $\mu \neq 75$, where $\bar{x} = 73$, $s = 15$, and $n = 100$. Use $\alpha = 0.05$.

7 Conduct the hypothesis test Ho: $\mu = 100$ versus Ha: $\mu > 100$, where $\bar{x} = 105$, $s = 30$, and $n = 10$. Use $\alpha = 0.10$.

8 Suppose that your critical value for a left-tailed hypothesis test is -1.96. For what values of the test statistic would you reject Ho?

Testing a Hypothesis about a Population Proportion

You use a *population proportion* hypothesis test when the variable is categorical (such as gender, political party, support/oppose, and so on) and you want to study only one population or group (such as all U.S. citizens or all registered voters). The test looks at the percentage (p) of individuals in the population that have a certain characteristic; for example, the percentage of households that have cellphones. The null hypothesis is Ho: $p = p_o$, where p_o is a certain value. For example, if the claim is that 20 percent of homes have cellphones, p_o is 0.20. The alternative hypothesis is one of the following: $p > p_o$, $p < p_o$, or $p \neq p_o$.

The formula for the test statistic for a single proportion is $\frac{\hat{p} - p_o}{\sqrt{\frac{p_o(1-p_o)}{n}}}$. To find it, follow these steps:

1. **Calculate the sample proportion, $\hat{p}$, by taking the number of people in the sample who have the characteristic of interest and dividing by n, the sample size.**
2. **Calculate the standard error, $\sqrt{\frac{p_o(1-p_o)}{n}}$. Save your answer.**
3. **Take the sample proportion minus p_o.**
4. **Divide your result from Step 3 by your result from Step 2.**

Compare your test statistic to the critical value that you previously set. If the test statistic is beyond the critical value(s), reject Ho.

See the following for an example of setting up and conducting a hypothesis test for a proportion.

EXAMPLE

Q. Suppose that a political candidate claims the percentage of uninsured drivers is 30 percent, but you believe the percentage is more than 30.

a. Set up your null and alternative hypotheses.

b. Find the critical value(s), assuming $\alpha = 0.05$ and you have a large sample size.

A. First, make sure you can identify that this is a hypothesis test about a proportion. It has a claim that's being challenged or tested, and the claim is about a percentage (or proportion). That's how you know.

a. The claim is that $p = 0.30$, so you put that into the null hypothesis. The alternative of interest is the one where the percentage is actually more (>) than 30. So you have Ho: $p = 0.30$ versus Ha: $p > 0.30$.

b. The critical value is $Z = 1.64$. Here's how you get it: Using the Z-table (see the Appendix), the area beyond the critical value must be 0.05. (In this case, *beyond* means "above," because you run a right-tailed test.) The Z-table uses percentiles; because 0.05 of the area is above your value, 0.95 must be below it. So look up the 95th percentile on the Z-table in the appendix to find the standard score. It is $Z = +1.6$. Or you can just use Table 13-1 because this problem uses 0.05, which just so happens to be included on it; you will also get $Z = +1.64$.

9 Carry out the hypothesis test of Ho: $p = 0.5$ versus Ha: $p > 0.5$, where $\hat{p} = 0.60$ and $n = 100$. Use $\alpha = 0.05$.

10 Carry out the hypothesis test of $p = 0.5$ versus $p < 0.5$, where $\hat{p} = 0.40$ and $n = 100$. Use $\alpha = 0.05$.

11 Carry out the hypothesis test of Ho: $p = 0.5$ versus Ha: $p \neq 0.5$, with $x = 40$ and $n = 100$, where x is the number of people in the sample that have the characteristic of interest. Use $\alpha = 0.01$.

12 Suppose that you want to test the fairness of a single die, so you concentrate on the proportion of 1s that come up. Write down the null and alternative hypotheses for this test.

Testing for a Difference between Two Population Means

This test is used when the variable is numerical (such as income, cholesterol level, or miles per gallon) and two populations or groups are being compared (such as men versus women, athletes versus nonathletes, or cars versus SUVs). Two separate random samples need to be selected, one from each population, to collect the data needed for this test. The null hypothesis is that the two population means are the same; in other words, their difference is equal to 0. The notation for the hypotheses is Ho: $\mu_x - \mu_y = 0$, where μ_x represents the mean of the first population, and μ_y represents the mean of the second population.

The formula for the test statistic comparing two means when both populations are normal (or approximately normal) and both population standard deviations are known is $\frac{(\bar{x} - \bar{y}) - 0}{\sqrt{\frac{\sigma_x^2}{n_1} + \frac{\sigma_y^2}{n_2}}}$.

To find it, follow these steps:

1. **Calculate the sample means, $\bar{x}$ and $\bar{y}$, and population standard deviations, σ_x and σ_y, for each sample separately. Let n_1 and n_2 represent the two sample sizes (they need not be equal).**

2. **Find the difference between the two sample means, $\bar{x} - \bar{y}$.**

3. **Calculate the standard error, $\sqrt{\frac{\sigma_x^2}{n_1} + \frac{\sigma_y^2}{n_2}}$. Save your answer.**

4. **Divide your result from Step 2 by your result from Step 3.**

Compare your test statistic to the critical value from the Z-distribution (Table 13-1).

TIP

If the sample sizes are small and you don't have a normal distribution and/or the population standard deviations are unknown, you substitute σ_x and σ_y with the sample standard deviations s_x and s_y and use the *t*-distribution with $n_1 + n_2 - 2$ degrees of freedom (see the *t*-table in the Appendix). If the test statistic is beyond the critical value(s), you reject Ho.

See the following for an example conducting a hypothesis test for two means.

EXAMPLE

Q. A teacher instructs two statistics classes with two different teaching methods (Group 1: computer versus Group 2: pencil/paper). She wants to see whether the computer method works better by comparing average final exam scores for the two groups. She selects volunteers to be in each group.

a. Has the teacher implemented a right-tailed, left-tailed, or two-tailed test?

b. The teacher doesn't try to control for other factors that can influence her results. Name some of the factors.

c. How can she change her study to improve the quality of her results?

A. The teacher uses a test for two population means, because she compares the average exam scores, and exam scores are a quantitative variable.

a. She wants to see whether the computer group does better, and she puts these students in Group 1, so she wants to show that the mean of Group 1 is greater than (>) Group 2. She implements a right-tailed test.

b. Some other factors include the intelligence level of the students, how comfortable they are with computers, the quality of the teaching activities, the way she writes the test, and so on.

c. She can improve her results by randomly assigning the students to groups, which creates a more level playing field, rather than asking for volunteers. She can also match up students according to the variables mentioned in part b and randomly assign one member of each pair to the computer group and the other to the pencil/paper group. This would require a paired *t*-test to do, which comes up later in this chapter.

13 **Conduct the hypothesis test Ho: $\mu_x - \mu_y = 0$ versus Ha: $\mu_x - \mu_y < 0$, where $\bar{x} = 7$, $\bar{y} = 8$, $\sigma_x = 2$, $\sigma_y = 2$, $n_1 = 30$, and $n_2 = 30$. Use $\alpha = 0.01$.**

14 **Conduct the hypothesis test Ho: $\mu_x - \mu_y = 0$ versus Ha: $\mu_x - \mu_y > 0$, where $\bar{x} = 75$, $\bar{y} = 70$, $\sigma_x = 15$, $\sigma_y = 10$, $n_1 = 50$, $n_2 = 60$. Use $\alpha = 0.05$.**

15 **Conduct the hypothesis test Ho: $\mu_x = \mu_y$ versus Ha: $\mu_x \neq \mu_y$, where $\bar{x} = 75$, $\bar{y} = 70$, $\sigma_x = 15$, $\sigma_y = 10$, $n_1 = 50$, $n_2 = 60$. Use $\alpha = 0.05$.**

16 **Suppose that you conducted a hypothesis test for two means (group one mean minus group two mean) and you reject Ho: $\mu_1 = \mu_2$ versus Ha: $\mu_1 \neq \mu_2$. You conclude that the two population means aren't equal. Can you say a little more? Explain how you can tell from the sign on the test statistic which group probably has the higher mean.**

Testing for a Mean Difference (Paired *t*-Test)

You use this test when the variable is numerical (such as income, cholesterol level, or miles per gallon) and when you pair up the individuals in the sample in some way (identical twins are often used) or use the same people twice (with a pre- and post-test, for example). Researchers typically use paired tests for medical studies when they test to see whether a certain treatment works, without having to worry about other factors associated with the subjects that may influence the results. For example, you want to compare a new blood pressure drug to an existing one to see whether it does a better job. To make it a fair test, you pair up the people in the study according to their weight, age, fitness level, and severity of blood pressure problems. With these pairing parameters, you can attribute any difference in blood pressure to the drug. (See Chapter 16 for more information on designed experiments like this one.)

You collect the data in pairs, and for each pair, you find the difference between the values. For example, suppose that you have a pair of subjects in the blood pressure test. Suppose that the first person in the pair is in the "current drug" group, and her blood pressure is 190 after the experiment. Suppose that the second person is in the "new drug" group, and her blood pressure is 180 after the experiment. The difference in blood pressures for this pair is $190-180$, or $+10$. (That means the new drug drops the blood 10 points more than the current drug for this pair.) You make the same calculations for each pair of subjects in the study.

The set of all the paired differences becomes your new (single) data set. This test is now the same as a test for a single population mean, and the null hypothesis is that the mean is equal to 0. (The average of all the differences should be 0 if the null hypothesis is true.) The notation for the null hypothesis is Ho: $\mu_d = 0$, where μ_d is the mean of the paired differences.

REMEMBER

The mean of the paired differences is different from the difference in the means. For the mean of the paired differences, you pair data, and you look at one set of differences for each pair and find the mean of that single data set. For the difference in the means, you have two different populations, so you have to compare their two means separately (see the section "Testing for a Difference between Two Population Means" earlier in this chapter).

The formula for the test statistic for paired differences is $\frac{\bar{d}-\mu_d}{\frac{s_d}{\sqrt{n}}}$. To calculate it, run through the following steps:

1. **For each pair of data, take the first value in the pair minus the second value to find the difference. Think of the differences as your new data set.**

2. **Calculate the mean $\bar{d}$ and the standard deviation (s_d) of all the differences. Let n represent the number of differences (or pairs) that you have (not necessarily the number of individuals).**

3. **Calculate the standard error, $\frac{s_d}{\sqrt{n}}$. Save your answer.**

4. **Find $\bar{d}-0$ divided by the standard error from Step 3.**

   Remember, $\mu_d = 0$ under the assumption that Ho is true.

If the number of pairs (n) is 30 or more, compare your test statistic to the critical value(s) from the standard normal distribution (see the Z-table in the Appendix or Table 13-1 earlier in this chapter). If the number of pairs (n) is less than 30, compare your test statistic to the critical value(s) from the t-distribution with $n-1$ degrees of freedom (see the t-table in the Appendix). If your test statistic is beyond the critical value(s), reject Ho. If not, fail to reject Ho.

See the following for an example involving a matched-pairs test.

EXAMPLE

Q. Suppose that you use a paired t-test using matched pairs to find out whether a certain weight-loss method works. You measure the participants' weights before and after the study, and you take weight before minus weight after as your pairs of data.

a. If you want to show that the program works, what's your Ha?

b. Explain why measuring the same participants both times rather than measuring two different groups of people (those on the program compared to those not) makes this study much more credible.

A. The signs on the differences are important when making comparisons. If you subtract two numbers and get a positive result, that means the first number is larger than the second. If the result is negative, the second is larger than the first.

a. If the program works, the weight loss (weight before minus weight after) has to be positive. So you have Ho: $\mu_d = 0$ versus Ha: $\mu_d > 0$.

b. If you use two different groups of people, you introduce other variables that can account for the weight differences. Using the same people cuts down on unwanted variability by controlling for possible confounding variables.

Note: If you switch the data around and take weight after minus weight before, you have to switch the sign in the alternative hypothesis to be < (less than). Most people like to use positive values, and greater-than signs (>) produce them. If you have a choice, always order the groups so the one that may have the higher average is Group 1.

17 Conduct the hypothesis test Ho: $\mu_d = 0$ versus Ha: $\mu_d > 0$, where $\bar{d} = 2$, $s_d = 5$, and $n = 10$. Use $\alpha = 0.05$.

18 Conduct the hypothesis test Ho: $\mu_d = 0$ versus Ha: $\mu_d > 0$, where $\bar{d} = 2$, $s_d = 5$, and $n = 30$. Use $\alpha = 0.05$.

Testing a Hypothesis about Two Population Proportions

You use this test when the variable is categorical (such as smoker/nonsmoker, political party, support/oppose an opinion, and so on) and when you want to know the percentage of individuals with a certain characteristic (such as the percentage of smokers). In this case, you compare two populations or groups (such as men versus women or Democrats versus Republicans). To conduct this test, you need to select two separate random samples — one from each population. The null hypothesis is that the two population proportions are the same; in other words, their difference is equal to 0. The notation for the null hypothesis is Ho: $p_1 - p_2 = 0$, where p_1 is the percentage from the first population and p_2 is the percentage from the second population.

The formula for the test statistic comparing two proportions is $\frac{(\hat{p}_1 - \hat{p}_2) - 0}{\sqrt{\hat{p}(1-\hat{p})\left(\frac{1}{n_1} + \frac{1}{n_2}\right)}}$. To find it, follow these steps:

1. **Calculate the sample proportions $\hat{p}_1$ and $\hat{p}_2$ for each sample. Let n_1 and n_2 represent the two sample sizes (they need not be equal).**
2. **Calculate the overall sample proportion, $\hat{p}$, which is the total number of individuals from both samples who have the characteristic of interest divided by the total number of individuals from both samples $(n_1 + n_2)$.**
3. **Find the difference between the two sample proportions, $\hat{p}_1 - \hat{p}_2$.**
4. **Calculate the standard error, $\sqrt{\hat{p}(1-\hat{p})\left(\frac{1}{n_1} + \frac{1}{n_2}\right)}$. Save your answer.**
5. **Divide your result from Step 3 by your result from Step 4.**

Compare your test statistic to the critical value that you previously set. If the test statistic is beyond the critical value(s), reject Ho. If not, fail to reject Ho. In most cases, the critical value will be on the Z-distribution because the sample sizes will be large in these situations. To be sure, check that $n\hat{p}$ and $n(1-\hat{p})$ are both at least 10.

See the following for an example of setting up a hypothesis test for two proportions.

EXAMPLE

Q. Suppose that you want to test whether there's a higher percentage of males who are Democrat than females who are Democrat.

a. Write down your null and alternative hypotheses.

b. Explain why it doesn't matter what the actual percentage of Democrats is for males or females.

A. Your two populations are males and females, and you compare the percentage of Democrats in each group. This means p_1 equals the percentage of all males who are Democrat, and p_2 equals the percentage of all females who are Democrat.

a. Your Ho is that the percentages are the same $(p_1 = p_2)$ versus the Ha that $p_1 > p_2$ (because you want to see whether the percentage of males is higher than the percentage of females).

b. Your only concern should be the difference in the percentage of Democrats for males and females and whether the difference is 0. A zero differential can happen in many ways; for example, both genders have about 40 percent Democrats, or both have 80 percent Democrats. The actual values of the proportions don't matter when you look at their differences.

Note: Saying Ho: $p_1 = p_2$ is the same as saying Ho: $p_1 - p_2 = 0$. Take the first equation and subtract p_2 from each side. The second version gives you a number to put in the null hypothesis (0), which is nice. That's because if the proportions are equal, their difference has to be 0.

19 Conduct the hypothesis test Ho: $p_1 - p_2 = 0$ versus Ha: $p_1 - p_2 > 0$, where $\hat{p}_1 = 0.60$, $\hat{p}_2 = 0.50$, $\hat{p} = 0.55$, $n_1 = 100$, $n_2 = 100$. Use $\alpha = 0.05$.

20 Conduct the hypothesis test Ho: $p_1 - p_2 = 0$ versus Ha: $p_1 - p_2 \neq 0$, where $x_1 = 1{,}000$, $x_2 = 1{,}100$, $n_1 = 2{,}500$, $n_2 = 2{,}500$. Use $\alpha = 0.05$.

Answers to Problems in Testing Hypotheses

1 You see $\overline{x}$ in both hypotheses, which is wrong. The null and alternative hypotheses make statements about the population parameter, not about the sample statistic.

WARNING

Instructors get really bent out of shape when they see this kind of mistake, so avoid it at all costs. Always keep sample statistics like $\overline{x}$ and $\hat{p}$ out of Ho and Ha. Use population parameters like μ and p instead.

2 The process of doing a hypothesis test in full requires several steps: setting up the hypotheses, finding the critical value, calculating the test statistic, and making the decision.

a. The claim is that the average pizza delivery time is 30 minutes, but you believe it takes longer than that. This means that Ho is $\mu = 30$ and Ha is $\mu > 30$.

b. The critical value is 1.96 looking at Table 13-1 because $\alpha = 0.05$ and it's a ">" alternative hypothesis, Ha.

c. Your sample of 10 pizzas has an average delivery time of 40 minutes. Assume that the population standard deviation is 15 minutes and the times have a normal distribution. That means the test statistic is $Z = \frac{\overline{x} - \mu_o}{\sigma/\sqrt{n}} = \frac{40 - 30}{15/\sqrt{10}} = 2.11$.

d. The conclusion is reject Ho because the test statistic 2.11 is > 1.96. The average pizza delivery time is likely to be more than 30 minutes based on these data.

TIP

Make sure you can identify these problems out of context — in other words, when the problems are all mixed up on an exam. You may want to copy some problems, mark where they come from, mix them up, and try to solve them. Also, as you go through the problems in this workbook, always think about how to recognize the types of problems in a test situation and how to approach them. I give you clues to look for; write them down in a quick outline to help you study.

3 Because the claim is that the average football game lasts 3 hours, and you believe it's more than that, the hypotheses are Ho: $\mu = 3$ and Ha: $\mu > 3$. Your random sample of 35 games has an average time of 3.25 hours, and the population standard deviation is 1 hour so the test statistic is $Z = \frac{\overline{x} - \mu_o}{\sigma/\sqrt{n}} = \frac{3.25 - 3}{1/\sqrt{35}} = 1.48$. Using $\alpha = 0.05$, the critical value is 1.96 from the Z-distribution so the conclusion is fail to reject Ho because the test statistic is less than the critical value. You can't say with these data that the average football game lasts more than 3 hours.

4 In a right-tailed test, Ha has a greater-than sign (>) in it. Your critical value on the Z-distribution is 1.64, because the probability of being below 1.64 is equal to 0.95 (see Table 13-1 earlier in this chapter), and the probability of being beyond (in this case, "above") 1.64 is $1 - .95 = .05$. If you run a left-tailed test (Ha has a less-than [<] sign in it), your critical value is $Z = -1.64$, because the area beyond (in this case, "below") −1.64 is 0.05. If you run a two-tailed test (Ha has a not-equal-to sign [≠] in it), you have two critical values — one positive and one negative — and the total area beyond them (in both directions) is 0.05. That means the area above the positive one is $.05 \div 2 = .025$, and the area below the negative one is 0.025. The critical values in that case are $Z = \pm 1.96$, as seen in Table 13-1.

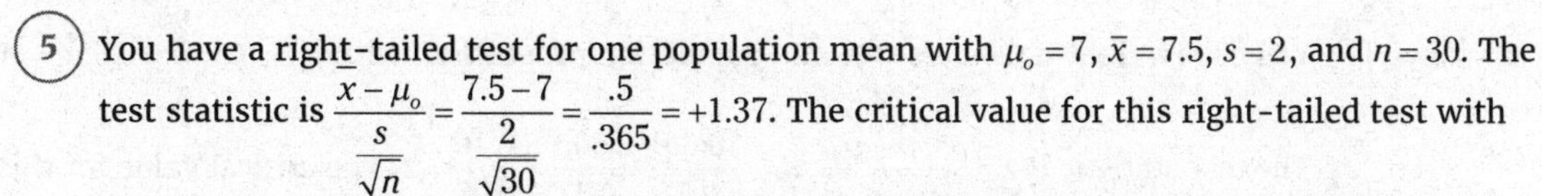

(5) You have a right-tailed test for one population mean with $\mu_o = 7$, $\bar{x} = 7.5$, $s = 2$, and $n = 30$. The test statistic is $\frac{\bar{x} - \mu_o}{\frac{s}{\sqrt{n}}} = \frac{7.5 - 7}{\frac{2}{\sqrt{30}}} = \frac{.5}{.365} = +1.37$. The critical value for this right-tailed test with $\alpha = 0.01$ is $+2.46$ (see t-table in the appendix with 29 degrees of freedom, column for .01.). *Conclusion:* You shouldn't reject Ho, because the test statistic is less than the critical value. *Interpretation:* According to your data, you don't have enough evidence to say the mean is more than 7.

TIP

Don't stop with the statistically correct conclusion: reject Ho or fail to reject Ho. Always go back to the question and try to answer it in the context of the problem as best as you can. Your instructor will love you for it.

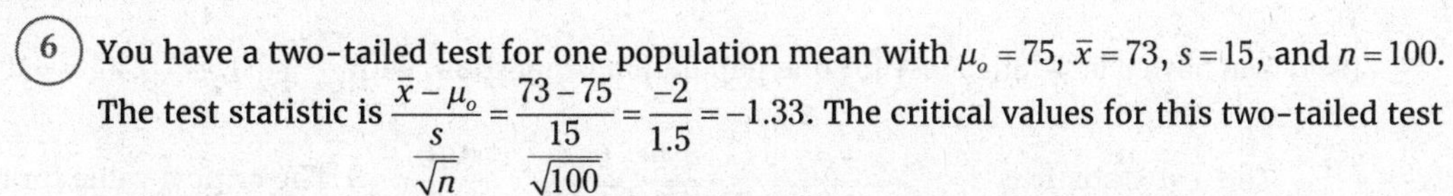

(6) You have a two-tailed test for one population mean with $\mu_o = 75$, $\bar{x} = 73$, $s = 15$, and $n = 100$. The test statistic is $\frac{\bar{x} - \mu_o}{\frac{s}{\sqrt{n}}} = \frac{73 - 75}{\frac{15}{\sqrt{100}}} = \frac{-2}{1.5} = -1.33$. The critical values for this two-tailed test with $\alpha = 0.05$ are -1.96 and $+1.96$ (see Table 13-1 since the t-distribution is about the same as the Z-distribution when n is 100.) *Conclusion:* You shouldn't reject Ho, because the test statistic is between rather than beyond the critical values. *Interpretation:* You don't have enough evidence to say that the mean for this population is anything but 75.

TIP

When doing problems involving hypothesis tests, I recommend you immediately write down what type of test you have and what the given information is, as I do in these solutions. It helps your instructor see where you're going, and it helps you keep it all straight.

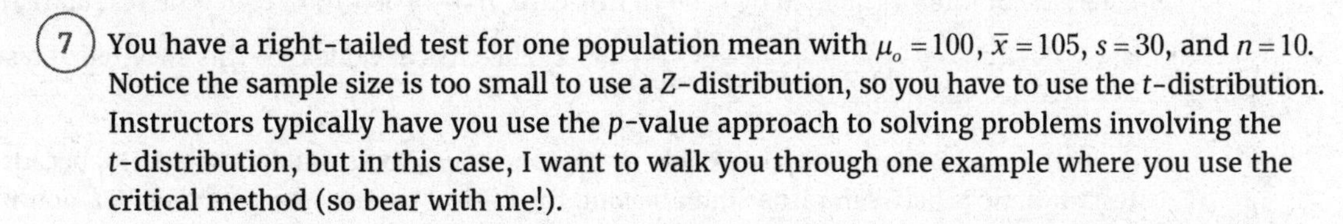

(7) You have a right-tailed test for one population mean with $\mu_o = 100$, $\bar{x} = 105$, $s = 30$, and $n = 10$. Notice the sample size is too small to use a Z-distribution, so you have to use the t-distribution. Instructors typically have you use the p-value approach to solving problems involving the t-distribution, but in this case, I want to walk you through one example where you use the critical method (so bear with me!).

The degree of freedom for your t-distribution is $n - 1 = 9$, so you can denote this as t_9. The test statistic is $\frac{\bar{x} - \mu_o}{\frac{s}{\sqrt{n}}} = \frac{105 - 100}{\frac{30}{\sqrt{10}}} = \frac{5}{9.487} = 0.53$. The critical value for this right-tailed test with $\alpha = .05$ is 1.833 on the t_9 distribution. Here's why: Because this is a right-tailed test, the area beyond (in this case, "above") the critical value must be $\alpha = 0.05$, which means the area above it must be .05. Find column .05 and the row for 9 degrees of freedom. You find 1.833, so you have your critical value (whew!). *Conclusion:* You shouldn't reject Ho, because the test statistic (0.53) isn't beyond the critical value. *Interpretation:* You don't have enough evidence to say that the mean is more than 100.

(8) Anything beyond the critical value leads to a rejection of Ho. In this case, the critical value is -1.96, so beyond means "less than"; therefore, any test statistic that comes in less than -1.96 leads you to reject Ho.

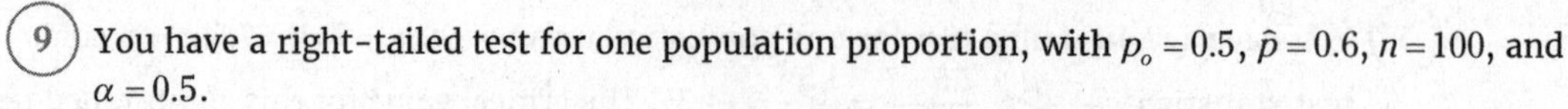

9. You have a right-tailed test for one population proportion, with $p_o = 0.5$, $\hat{p} = 0.6$, $n = 100$, and $\alpha = 0.5$.

The test statistic is $z = \dfrac{\hat{p} - p_o}{\sqrt{\dfrac{p_o(1-p_o)}{n}}} = \dfrac{0.60 - 0.50}{\sqrt{\dfrac{0.5(1-0.5)}{100}}} = \dfrac{0.10}{0.05} = 2$. The critical value for this right-tailed test with $\alpha = 0.05$ is $+1.64$ (see Table 13-1). *Conclusion:* You should reject Ho, because the test statistic is beyond the critical value. *Interpretation:* According to your data, the proportion of people in the population who have the characteristic of interest is more than 0.50.

TIP

Always use the decimal version of p when you work with hypothesis tests for one or two proportions. The formulas don't work if you use percents.

10. You have a left-tailed test for one population proportion, with $p_o = 0.5$, $\hat{p} = 0.40$, $n = 100$, and $\alpha = 0.05$.

The test statistic is $z = \dfrac{\hat{p} - p_o}{\sqrt{\dfrac{p_o(1-p_o)}{n}}} = \dfrac{0.40 - 0.50}{\sqrt{\dfrac{0.5(1-0.5)}{100}}} = \dfrac{-0.10}{0.05} = -2$. The critical value for this left-tailed test with $\alpha = 0.05$ is -1.64 (see Table 13-1). *Conclusion:* You should reject Ho, because the test statistic is beyond the critical value. *Interpretation:* According to your data, the proportion of people in the population who have the characteristic of interest is less than 0.5.

11. You have a two-tailed test for one population proportion, with $p_o = 0.5$, $x = 40$, $n = 100$, and $\alpha = 0.01$. The sample proportion, $\hat{p}$, is the number of people in the sample who have the characteristic of interest divided by n, so in this case, $\hat{p} = \dfrac{x}{n} = 40 \div 100 = 0.40$. The test statistic is $z = \dfrac{\hat{p} - p_o}{\sqrt{\dfrac{p_o(1-p_o)}{n}}} = \dfrac{0.40 - 0.50}{\sqrt{\dfrac{0.50(1-0.50)}{100}}} = \dfrac{-0.10}{0.05} = -2$. The critical values for this two-tailed test with $\alpha = 0.01$ are -2.58 and $+2.58$ (see Table 13-1). *Conclusion:* You shouldn't reject Ho, because the test statistic is between rather than beyond the critical values. *Interpretation:* You don't have enough evidence to say that the proportion of this population who fall in the group of interest is anything but 0.50.

12. If the die is fair, each face should show up one-sixth of the time. If you let p equal the proportion of times this die will show a 1, you have Ho: $p = 1/6$. The alternative is that the coin isn't fair, so either $p < 1/6$ or $p > 1/6$ fit that case, which means you have Ha: $p \neq 1/6$.

13. You have a left-tailed test for two population means with $\bar{x} = 7$, $\bar{y} = 8$, $\sigma_x = 2$, $\sigma_y = 2$, $n_1 = 30$, $n_2 = 30$, and $\alpha = 0.01$. The test statistic is $\dfrac{(\bar{x} - \bar{y}) - 0}{\sqrt{\dfrac{\sigma_x^2}{n_1} + \dfrac{\sigma_y^2}{n_2}}} = \dfrac{(7-8) - 0}{\sqrt{\dfrac{2^2}{30} + \dfrac{2^2}{30}}} = \dfrac{-1}{0.516} = -1.94$. The critical value for this left-tailed test with $\alpha = 0.01$ is -2.33 (see Table 13-1). *Conclusion:* You shouldn't reject Ho, because the test statistic isn't beyond the critical value. *Interpretation:* According to your data, you can't say the difference in the means of these two populations is less than 0 (indicating no statistically significant difference between the means of these two populations).

REMEMBER

If you use $\alpha = 0.05$ in Question 13, the critical value is -1.96, and the test statistic of -1.94 is very close to this. You still can't reject Ho, but you can argue that the close number is a marginal result. The p-value approach (Chapter 14) helps solve the problem of getting different conclusions when you use different α levels. It allows you to report the strength of your test statistic and to let other people decide for themselves whether the info is enough to reject Ho.

14. You have a right-tailed test for two population means, with $\bar{x}=75$, $\bar{y}=70$, $\sigma_x=15$, $\sigma_y=10$, $n_1=50$ $n_2=60$, and $\alpha=0.05$.

The test statistic is $\frac{(\bar{x}-\bar{y})-0}{\sqrt{\frac{\sigma_x^2}{n_1}+\frac{\sigma_y^2}{n_2}}}=\frac{(75-70)-0}{\sqrt{\frac{15^2}{50}+\frac{10^2}{60}}}=\frac{5}{2.48}=2.02$. The critical value for this right-tailed test with $\alpha=0.05$ is +1.64 (see Table 13-1). *Conclusion:* You should reject Ho, because the test statistic is beyond the critical value. *Interpretation:* According to your data, the difference in the means of these two populations is greater than 0 (indicating a statistically significant difference between the means of these two populations).

15. This is the same hypothesis test as Question 14, except you run a two-tailed test rather than a right-tailed test. Notice that Ho and Ha appear different from usual in this problem, but notice that Ho: $\mu_x=\mu_y$ is the same as Ho: $\mu_x-\mu_y=0$ (just subtract μ_y from each side of the original equation). This is another way your professor may write these hypotheses, so be ready for it.

To work the problem, note that, again, you have a two-tailed test for two population means, with $\bar{x}=75$, $\bar{y}=70$, $\sigma_x=15$, $\sigma_y=10$, $n_1=50$, $n_2=60$, and $\alpha=0.05$. Again, the test statistic is 2.02. However, the critical values for this (now) two-tailed test with $\alpha=0.05$ are −1.96 and +1.96 (see Table 13-1). *Conclusion:* You should still reject Ho, because the test statistic is beyond the critical value, but it isn't as far beyond the critical value as it is in Question 14. *Interpretation:* According to your data, the difference in the means of these two populations is greater than 0, indicating a statistically significant difference between the means of these two populations. (And because the difference is positive, you can say that the first population has a larger mean than the second population.)

REMEMBER

Two-tailed tests require you to split the α level in half. The half level means the probability of being outside the critical values is cut in half for each side, making the critical values push out farther on each edge. In other words, if you can make a goal by going to either end of the field, you have to push both goal posts out farther to make up for it. So the result you have with a one-tailed test isn't as strong when you go to the two-tailed test. A one-tailed test puts all your eggs in one basket. A two-tailed test makes you divide your eggs into two baskets, so to speak.

16. After Ho has been rejected and you conclude that the groups don't have the same mean, the numerator of your test statistic $\bar{x}-\bar{y}$ tells you which group had the larger mean. If the numerator of your test statistic is positive, $\bar{x}-\bar{y}>0$, which means $\bar{x}>\bar{y}$. That means in terms of the populations, the mean of group one (which is μ_1) is likely to be larger than the mean of group two (which is μ_2). If the numerator of your test statistic is negative, $\bar{x}-\bar{y}<0$, which means $\bar{x}<\bar{y}$. That means in terms of the populations, the mean of group one (which is μ_1) is likely to be smaller than the mean of group two (which is μ_2). Of course, these results all depend on the samples being representative of their populations.

17. You have a right-tailed test for the average difference with $\bar{d}=2$, $s_d=5$, $n=10$, and $\alpha=0.05$. You do this test the same way you conduct a test for one population mean. The test statistic is $\frac{\bar{d}-0}{\frac{s_d}{\sqrt{n}}}=\frac{2-0}{\frac{5}{\sqrt{10}}}=\frac{2}{1.58}=1.27$. Because the sample size is under 30, you compare your test statistic to the *t*-distribution with $n-1=9$ degrees of freedom, denoted t_9. On that distribution, 1.27 falls between column .25 and .10. The *p*-value falls between these numbers. *Conclusion:* You shouldn't reject Ho, because the *p*-value isn't smaller than .05. *Interpretation:* According to your data, you can't say the mean difference for this population is greater than 0 (indicating no statistically significant mean difference for this population).

18 You have the same hypothesis test as Question 17 here, except the sample size is larger for this test. You have a right-tailed test for the average difference with $\bar{d}=2$, $s_d=5$, $n=30$, and $\alpha=0.05$.

The test statistic is $\frac{\bar{d}-0}{\frac{s_d}{\sqrt{n}}}=\frac{2-0}{\frac{5}{\sqrt{30}}}=\frac{2}{.913}=2.19$, which is larger than the test statistic in Question 17.

Because the sample size is 30 or more, you compare your test statistic to the *Z*-distribution. The critical value for this right-tailed test with $\alpha=0.05$ is 1.64. *Conclusion:* You should reject Ho, because the test statistic is beyond the critical value. *Interpretation:* According to your data, the mean difference for this population is greater than 0 (indicating a statistically significant and positive mean difference for this population).

TIP

A larger sample size makes it easier to reject Ho. It makes your test statistic more extreme, which increases its chances of crossing over into the rejection region. And in the case of Question 18, where the sample size goes from 10 to 30, the critical value decreases, because you don't have to use the *t*-distribution anymore. Remember, the *t*-distribution makes you pay a penalty for having less data, and that penalty involves pushing out the tails of the *t*-distribution so you have to get farther out there to reject Ho. A wider boundary makes it harder to reject Ho. (See Chapter 8 for more on the *t*-distribution.)

19 You have a right-tailed test for two population proportions with $\hat{p}_1=.60$, $\hat{p}_2=.50$, $\hat{p}=.55$, $n_1=100$, $n_2=100$, $\alpha=0.05$.

The test statistic is $\frac{(\hat{p}_1-\hat{p}_2)-0}{\sqrt{\hat{p}(1-\hat{p})\left(\frac{1}{n_1}+\frac{1}{n_2}\right)}}=\frac{(0.60-0.50)-0}{\sqrt{0.55(1-0.55)\left(\frac{1}{100}+\frac{1}{100}\right)}}=\frac{0.10}{0.07}=+1.43$.

The critical value for this right-tailed test with $\alpha=0.05$ is +1.64 (see Table 13-1). *Conclusion:* You shouldn't reject Ho, because the test statistic isn't beyond the critical value. *Interpretation:* According to your data, we can't say the difference in the proportions for these two populations is greater than 0 (indicating no statistically significant difference between the proportions for these two populations).

20 You have a two-tailed test for two population proportions with $\hat{p}_1=\frac{x_1}{n_1}=\frac{1{,}000}{2{,}500}=0.40$, $\hat{p}_2=\frac{x_2}{n_2}=\frac{1{,}100}{2{,}500}=0.44$, $\hat{p}=\frac{x_1+x_2}{n_1+n_2}=\frac{1{,}000+1{,}100}{2{,}500+2{,}500}=\frac{2{,}100}{5{,}000}=0.42$. The test statistic is

$$\frac{(\hat{p}_1-\hat{p}_2)-0}{\sqrt{\hat{p}(1-\hat{p})\left(\frac{1}{n_1}+\frac{1}{n_2}\right)}}=\frac{(0.40-0.44)-0}{\sqrt{0.42(1-0.42)\left(\frac{1}{2{,}500}+\frac{1}{2{,}500}\right)}}=\frac{-0.04}{0.0140}=-2.86.$$

The critical values for this two-tailed test with $\alpha=0.05$ are ±1.96 (see Table 13-1). *Conclusion:* You should reject Ho, because the test statistic is beyond the critical values. *Interpretation:* According to your data, the difference in the proportions for these two populations isn't equal to 0, indicating a statistically significant difference between the proportions for these two populations. (And because the test statistic is negative [taking Group 1 – Group 2], the proportion in the first population who have that characteristic of interest is likely to be lower than the proportion in the second population who have the characteristic of interest.)

» Finding *p*-values

» Examining *p*-values

» Keeping Type I and Type II errors straight

Chapter 14
Taking the Guesswork Out of p-Values and Type I and II Errors

So you've battled your way through hypothesis testing: null and alternative hypotheses, test statistics, critical values, rejecting or not rejecting Ho, and all that. You feel you finally have it all figured out when, suddenly, out of nowhere comes this thing called a *p*-value, and it rocks your world. Not to worry! This chapter helps you set the boat upright again for smooth sailing. This chapter helps you after you find your test statistics (see Chapter 11) and you want to make your conclusions correctly by using *p*-values. It also helps you interpret *p*-values that you see on homework, exams, or in the real world. (Did I say exams are the real world?)

Just a quick review before you start. Every hypothesis test has two hypotheses: the null hypothesis, Ho, and the alternative hypothesis, Ha. You cannot go against Ho unless you have sufficient evidence to provide reasonable doubt for Ho (just like in a jury trial). If you have sufficient evidence against Ho, you reject it in favor of the alternative hypothesis. Now how much evidence is needed to provide that level of reasonable doubt? In a trial, that value is nearly impossible to quantify, but luckily in statistics, things are more cut-and-dried. That's where the *p*-value really comes in handy.

Understanding What p-Values Measure

You use p-values to make a decision about a hypothesis test. They go beyond saying, "I reject the null hypothesis in favor of the alternative hypothesis," by giving you a specific number to indicate how much evidence you actually have against the null hypothesis. Do you have plenty of evidence against the null hypothesis? If so, you have a small p-value. Does your evidence give more support to the null hypothesis? If so, you have a larger p-value. Where's the cutoff point? For most cases, it's 5 percent. You also use p-values to indicate how strong your results are, allowing observers to interpret them on their own.

After you have your test statistic from your hypothesis test (see Chapter 13), you look it up on the standard normal (or Z) distribution (or the t-distribution if your sample size is small — less than 30; see Chapter 8) to find the probability of being at that value or beyond it. The probability shows you how likely it is that you'll get the test statistic if the claim is actually true. The farther out your test statistic is on the tails of the distribution, the less evidence you have for the null hypothesis.

The probability of being at or beyond your test statistic is called the *p-value,* and it represents the chance that your sample results could have occurred when Ho (the claim) is actually true. Small p-values mean the claim is going to be rejected. Large p-values mean the claim can't be rejected. I like to think of p-values as the strength of your evidence against Ho.

See the following for an example of interpreting a p-value.

EXAMPLE

Q. Suppose that you hear a researcher say she found a "statistically significant result," and in parentheses you see ($p = 0.001$). Explain what "statistically significant result" must mean.

A. The $p = 0.001$ part means that the p-value is 0.001, which is very small by most researchers' standards. Small p-values mean you reject the null hypothesis. A statistically significant result means the researcher has enough evidence to reject the null hypothesis when she conducts her hypothesis test.

Bob and Teresa each collect their own samples to test the same hypothesis. Bob's p-value turns out to be 0.05, and Teresa's turns out to be 0.01.

a. Why don't Bob and Teresa get the same p-values?

b. Who has stronger evidence against the null hypothesis: Bob or Teresa?

Does a small p-value mean that the null hypothesis isn't true? Explain.

Test (Statistic) Time: Figuring Out p-Values

To find the p-value for your test statistic, you

1. **Look up the location of your test statistic on the appropriate distribution — the standard normal (Z) distribution or the t-distribution.**

   You can find both in the Appendix.

2. **Find the probability of being at or *beyond* your test statistic in the same direction.**

   If the test statistic is positive, you want the probability of being greater than your test statistic. If the test statistic is negative, you want the probability of being less than your test statistic.

TIP

These rules for finding the p-value apply in 99.9 percent of all situations when Ha has a particular direction. The only remaining situations are those that instructors might come up with on an exam just to be — dare I say — tricky. Those are the never-found-in-the-real-world cases, where your Ha says $>$ and your test statistic is negative, or where your Ha says $<$ and your test statistic is positive. (People doing studies generally know what direction they expect their test statistic to fall and are seldom totally wrong about it.) In both of these cases, you know your p-value will be very large (greater than 0.50) because the test statistic is in the totally opposite direction of what you are testing for. So you don't have to do any work. Just say that it fails to reject Ho.

3. **Double the probability you find if (and only if) Ha is the "not equal to" alternative.**

   This accounts for both the "less than" and the "greater than" possibilities before the fact.

See the following for an example of finding a p-value for a given test statistic.

EXAMPLE

Q. Suppose that you want to test Ho: $\mu = 0$ versus Ha: $\mu > 0$, and your test statistic is $Z = 1.96$ (based on a large sample). What's your p-value?

A. Because the alternative hypothesis is ">," you need to find the area above 1.96. Find the standard score closest to $Z = 1.96$ on the Z-table (see the Appendix). The corresponding percentile is 97.50 percent, or 0.9750, which means the area beyond (in this case, above) 1.96 is $1 - 0.9750 = 0.025$. This is the p-value. (When you work with p-values, use them in their decimal form, not as percents.)

3 Suppose that you want to test Ho: $\mu = 0$ versus Ha: $\mu < 0$, and your test statistic is $Z = -2.5$. (Assume a large sample.) What's your p-value?

4 Suppose that you want to test Ho: $p = ½$ versus Ha: $p \neq ½$, and your test statistic is $Z = 0.5$. (Assume a large sample.) What's your p-value?

5 Suppose that you want to test Ho: $p = ½$ versus Ha: $p \neq ½$, and your test statistic is $Z = -1.2$. (Assume a large sample.) What's your p-value?

6 Suppose that you want to test Ho: $\mu = 0$ versus Ha: $\mu > 0$, and your test statistic (based on a sample of 10) is $t = 1.96$. What's your p-value?

7 Suppose that you want to test Ho: $\mu = 0$ versus Ha: $\mu < 0$, and your test statistic (based on a sample of 10) is $t = -2.5$. What's your p-value?

8 Suppose that you want to test Ho: $\mu = 0$ versus Ha: $\mu > 0$, and your p-value is 0.0446. (Assume a large sample.) What's the test statistic?

The Value Breakdown: Interpreting p-Values Properly

To interpret a p-value, size matters:

- For small p-values (generally less than 0.05), reject Ho. Your data doesn't support Ho, and your evidence is beyond a reasonable doubt.
- For large p-values (generally greater than 0.05), fail to reject Ho. Your sample data doesn't contain enough evidence to refute it.
- For p-values on or close to the borderline between accepting and rejecting, you produce "marginal" results (that can go either way).

TIP

Although some statistics instructors can go either way on the rejection issue, you should stay on the statistically correct and safe side: say "fail to reject Ho" rather than "accept Ho." Just trust me on this one.

Most statisticians use 5 percent as the cutoff for rejecting Ho and choosing Ha. Some people may have stricter cutoffs, such as 0.01, that require more evidence. Each reader makes his or her own decision about personal cutoff points, which is why researchers should report p-values rather than just personal reject/don't reject conclusions based on their own predetermined cutoffs. Statisticians call these preset cutoff probabilities *alpha levels*, denoted by α. (That's alpha, not a fish!) Typical values for α are 0.05 or 0.01, but instructors may use others (which may appear on exams).

Here's how to interpret a p-value for any given alpha level α:

- If the p-value is greater than or equal to α, you fail to reject Ho.
- If the p-value is less than α, reject Ho.
- If the p-value is on the borderline (very close to α), treat it as a marginal result.

TIP

You may be wondering how to interpret p-values that are close to the actual value of alpha. Here are some general rules for making conclusions based on p-values, assuming you have the most common alpha level of all, $\alpha = 0.05$:

- If the p-value is less than 0.001 (very small), the results are usually deemed "highly statistically significant."
- If the p-value is between 0.05 and 0.001 (small), the results are usually deemed "statistically significant."
- If the p-value is close to 0.05, the results are usually deemed "marginally significant."
- If the p-value is greater than (but not close to) 0.05, the results are usually deemed nonsignificant. (***Note:*** Statisticians don't like to say "insignificant" because nonsignificant results are still important, and saying "insignificant" would imply that they're not.)

See the following for an example of making proper conclusions given a p-value.

EXAMPLE

Q. Suppose that you're doing a statistical study and you find a p-value of 0.026 when you test Ho: $p = 0.25$ versus Ha: $p < 0.25$. You report this result to your fellow researchers.

a. What would a fellow researcher with $\alpha = .05$ conclude?

b. What would a fellow researcher with $\alpha = 0.01$ conclude?

A. Always compare the p-value to the alpha level chosen beforehand, using the rules mentioned earlier.

a. Because the p-value 0.026 is less than 0.05, this researcher has enough evidence to reject Ho.

b. Because the p-value 0.026 is greater than 0.01, this researcher doesn't have enough evidence to reject Ho.

9 Suppose that you want to test Ho: $\mu = 0$ versus Ha: $\mu < 0$, and your p-value turns out to be 0.06. What do you conclude? Use $\alpha = 0.05$.

10 Suppose that you want to test Ho: $\mu = 0$ versus Ha: $\mu < 0$, and your p-value turns out to be 0.60. What do you conclude? Use $\alpha = 0.05$.

11 Suppose that you want to test Ho: $\mu = 0$ versus Ha: $\mu < 0$, and your p-value turns out to be 0.05. What do you conclude? Use $\alpha = 0.05$.

12 Suppose that you want to test Ho: $p = \frac{1}{2}$ versus Ha: $p \neq \frac{1}{2}$, and your p-value turns out to be 0.95. What do you conclude? Use $\alpha = 0.05$.

Deciphering Type I Errors

Suppose a company claims that its average package delivery time is two days. A consumer group tests this hypothesis and concludes that the claim is false and that the average delivery time is actually more than two days. This is a big deal! If the group of researchers can stand by their statistics, they've done well to inform the public about the false advertising issue. But what if their data is wrong? Can this happen?

The answer is yes. Why? Because they base the conclusions on a sample of packages, not the entire population, the conclusions based on their particular sample may lead to the wrong conclusion. Your sample, although you collect it randomly, may be an atypical sample whose result ends up far out on the distribution. And Ho could be true, but your results lead you to a different conclusion.

Incorrectly rejecting Ho is called a *Type I error*. I liken it to a false alarm. In the case of the packages, if the consumer group makes a Type I error when it rejects the company's claim, it issues a false alarm. What's the result? A very angry packaging company, I guarantee that!

How often does a Type I error happen? Whatever your given cutoff probability is for rejecting Ho. The probability of a Type I error is the same as the alpha level in a hypothesis test. For example, if the alpha is 0.05, the probability of a Type I error is 0.05.

So what do you need to know about Type I errors? The most important details to know are the definition of Type I error; that a Type I error should be small (so choose a small alpha); how to describe one in the context of a problem; and the consequences of making a Type I error, such as upsetting the people who made the claim in Ho to begin with, because they were right and you concluded they were not.

See the following for an example of understanding when a Type I error is possible.

Q. Suppose that you want to test Ho: $\mu = 0$ versus Ha: $\mu < 0$, and your p-value turns out to be 0.006. You reject Ho. (Your alpha level was 0.05.) Explain what a Type I error would mean in this situation if such an error is possible.

A. For a Type I error to happen, Ho had to be true, and you rejected it. This could have happened here, because you did reject Ho. No one actually knows whether Ho is true, but if it is, you made a Type I error. That would mean trouble for you, because you rejected a claim that turned out to be true. *Note:* The probability of a Type I error here is 0.05, not 0.006.

13 Suppose that you want to test Ho: $\mu = 0$ versus Ha: $\mu < 0$, and your *p*-value turns out to be 0.6. You fail to reject Ho. Explain why you couldn't have made a Type I error here.

14 Suppose a gubernatorial candidate claims that she would get 60 percent of the vote if the state held the election today. You test her hypothesis, believing that the actual percentage is less than 60. Suppose you make a Type I error.

a. Did you reject Ho or not?

b. Is Ho true or not?

c. Describe the impact of making a Type I error in this situation.

Deciphering and Distinguishing Type II Errors

Suppose that a packing company promising 2-day delivery isn't delivering on its claim, hauling around late shipments. Can you say that a consumer group's testing will for sure detect this problem? If the actual delivery time is 2.1 days rather than 2 days, the group may not detect it, but if the actual delivery time is 3 days, a consumer group doesn't have to take a very big sample to figure out something's up. The issue lies with the in-between values, such as 2.5 days. Calculating a Type II error is beyond the scope of an introductory statistics class, but you can still understand and interpret a Type II error.

Not rejecting Ho when you should is called a *Type II error*. Some statisticians describe the error as the probability of you failing to reject Ho when Ha is true. I like to call it a "missed detection."

In the case of the packages, if the company doesn't deliver on its claim and you don't detect the tardiness, the company can go on misleading the public and disappointing customers — the consequences of making a Type II error in this case.

REMEMBER

Type II errors are more complicated than Type I errors because they depend on what the actual true value of the population mean or proportion was, not just the direction of Ha. For most introductory statistics courses, finding a Type II error is beyond the scope of the class, as is such with this workbook. So what do you need to know about Type II errors? The most important details to know are the definition of a Type II error; that the probability of it should be small; how to describe one in the context of a problem; and that the consequences of making one are bad for you because you missed an opportunity to blow the whistle on a false claim, and you didn't do it. And you need to be able to distinguish Type II errors from Type I errors.

By trying to make a Type II error smaller, you may make the Type I one bigger, and vice versa. Think of a jury trial situation. If you want to make sure criminals don't get away with crimes, you make it easier to convict and increase the risk of innocent people going to jail. If you want to make sure no innocent people go to jail, you make it harder to convict and increase the risk of more criminals getting away with their crimes. Bigger samples generally reduce Type II errors because more data helps you do a better job of detecting even small deviations from the null hypothesis. And small alpha levels generally reduce Type I errors because they reduce the chance of you rejecting Ho in the first place.

See the following for an example of how to distinguish Type I errors from Type II errors.

EXAMPLE

Q. In the jury trial example in the previous paragraph, which error in the decision made by the jury is the Type I error and which is the Type II error?

A. A Type I error means a false alarm; you rejected Ho when you didn't need to. That's the case where you sent an innocent person to jail. The Type II error is the missed detection. You should have rejected Ho, but you didn't. That means you let a guilty person go free.

15 Suppose that you test to see if cereal boxes are underfilled (Ho: $\mu = 18$ ounces versus Ha: $\mu < 18$ ounces). Describe what Type I and Type II errors would mean in this situation.

16 Can you make a Type I error and a Type II error at the same time?

17 Suppose that your friend claims that a coin is fair, but you don't think so. You decide to test it. Your p-value is 0.045 (at alpha level 0.05). You reject Ho: $p = \frac{1}{2}$. Suppose that the coin really is fair.

a. Which type of error did you make?

b. Describe the impact of your error.

18 Suppose that you want to test whether or not a machine makes its widgets according to specifications by taking a sample. Ho says the machine is fine, and the alpha is 0.05. Your p-value is 0.3, so you decide the machine works to specification.

a. Suppose that the machine really doesn't work properly. Which type of error did you make?

b. What's the impact of your error?

Answers to Problems in p-Values and Type I and II Errors

1. Remember that because sample results vary, so do their p-values.

   a. Bob and Teresa take different samples, which leads to different test statistics, which gives different percentiles from the Z-table (found in the Appendix). The different percentiles lead to different p-values.

   b. Teresa's p-value is smaller, so her evidence against Ho is stronger.

REMEMBER

Always think of a p-value as the strength of the evidence against the null hypothesis. The null hypothesis is on trial, and the p-value is the prosecutor.

2. No, it just means that you have plenty of evidence against the null hypothesis. The null hypothesis can still be true. You never know what the truth is; the best you can do is to make a conclusion based on your sample.

3. Find $Z = -2.5$ on the Z-table (see the Appendix). The corresponding percentile is 0.62 percent, or 0.0062, which means the area beyond (in this case, below) -2.5 is 0.0062. (You don't subtract from 1 because the percentile already gives you the area below, and when the test statistic is negative, that's what you want.) The p-value is 0.0062.

TIP

p-values are probabilities and must be between 0 and 1. If you find a p-value that's greater than 1 or less than 0, you know that you've made a mistake.

4. Find $Z = 0.5$ on the Z-table (see the Appendix). The corresponding percentile is 69.15 percent, or 0.6915, which means the area beyond (in this case, above, because 0.5 is positive) 0.5 is $1 - .6915 = .3085$. The p-value is $.3085 * 2 = .617$. (Don't forget to double the p-value if Ha has "not equal to" in it.)

5. Find $Z = -1.2$ on the Z-table (see the Appendix). The corresponding percentile is 11.51 percent, or 0.1151. (You take this probability as it is because area "beyond" a negative number means area below it.) The p-value is $.1151 * 2 = .2302$. Again, you double the value because of the two-sided (not equal to) alternative hypothesis (see Chapter 13).

6. Here, $t = 1.96$, and the sample size is $n = 10$, so you have to look at a t-distribution with $10 - 1 = 9$ degrees of freedom. Use the table in the Appendix to look at the row corresponding to 9 degrees of freedom. Your test statistic, 1.96, falls between the columns for 0.05 and 0.025. All you can say is $0.025 < p\text{-value} < 0.05$.

REMEMBER

Computer software can give you an exact p-value for any test statistic. However, if you calculate p-values by using tables, and you need to use a t-table, chances are you have to settle for a p-value falling between two numbers.

7. Here, $t = -2.5$, and the sample size is $n = 10$, so you have to look at a t-distribution with $10 - 1 = 9$ degrees of freedom. Look at the row corresponding to 9 degrees of freedom on the table in the Appendix. Negative test statistics don't appear on the t-table, so you have to use symmetry to get your answer. Look for $+2.5$. Notice that it falls between the columns for 0.025 and 0.01. In other words, $0.01 < p\text{-value} < 0.025$.

8. Your p-value is 0.0446, or 4.46 percent, so by looking at the Z-table (in the Appendix), the standard score (Z-value) corresponding to a percentile of 4.46 percent is -1.7. However, your Ha has a ">" sign, indicating that your test statistic is a positive value. So your test statistic is actually $+1.7$. ***Note:*** The probability of being beyond 1.7 is 100 percent minus the percentile that goes with 1.7, which is 95.54 percent. The symmetry shouldn't surprise you: The probability equals 4.46 percent.

9. Because the p-value 0.06 is greater than $\alpha = 0.05$, you don't have quite enough evidence to reject Ho. But you could say that these results are marginally significant.

10. Because the p-value 0.6 is greater than $\alpha = 0.05$, you don't have nearly enough evidence to reject Ho. In other words, the results are nonsignificant.

11. This situation resembles a flipped coin that lands on its side; it can go either way. Because the p-value is exactly equal to 0.05, you can say the results are marginal. Reporting the p-value is important, so in borderline situations like this, everyone can make a personal decision. (Of course, you can play it like a true statistician and flip a coin to tell you what to do.) But in this situation, you might want to check with your instructor and see what he/she wants you do to.

12. Large p-values, such as 0.95, always result in failing to reject Ho. Make sure you don't get 0.95 (from confidence intervals; see Chapter 11) and 0.05 mixed up and make the wrong conclusion. You need small p-values to reject Ho, and 0.95 isn't a small p-value.

13. You can't make a Type I error in this situation, because to make a Type I error, you have to reject Ho. You didn't. If Ho really is true in this case, your conclusion is correct.

14. Sometimes when you make a decision, you can be wrong, and if you are, your mistake makes some kind of impact.

    a. If you make a Type I error, you reject Ho, so yes.

    b. If you make a Type I error, that assumes Ho is true.

    c. The impact is that you conclude she can't get the votes, when in reality she can. If you're her campaign manager, you should probably look for another job.

15. A Type I error means you make a false alarm. You reject Ho (the boxes are fine) and conclude Ha (the boxes are underfilled), when in truth, the cereal boxes are just fine. You may get into trouble with the cereal company if you press your point. A Type II error means you miss detection. You fail to reject Ho when you really should have, because Ha is true: The boxes are underfilled. You let the company get by with cheating its customers.

16. In any given situation, you can make only one of the two errors; you can't make both at the same time. Type I errors can happen only when Ho is true, and Type II errors can happen only when Ho is false (in other words, when Ha is true). Both can't happen at the same time. The problem is, you typically can't identify the true situation, so you have to be prepared to identify and discuss both types of errors.

17 With a problem like this, it helps to first write down Ho and Ha, even if you put them in words. Here, you have Ho: coin is fair versus Ha: coin isn't fair. (Always put the "status quo" or "innocent until proven guilty" item in the null hypothesis.)

a. Your *p*-value suggests that the coin isn't fair, so you reject Ho. But the coin really is fair, so Ho turns out to be true. Rejecting Ho when Ho is true is a Type I error. You're guilty of making a false alarm. Do you plead the fifth?

b. Your friend may be pretty upset with you, or you may be the subject of subsequent ridicule, I guess.

TIP

A Type I error is a false alarm, which is much easier to remember than "you reject Ho when Ho is true."

18 First, you write down Ho and Ha, even if you put them in words. Here, you have Ho: machine is okay versus Ha: machine isn't okay. (Always put the "status quo" or "innocent until proven guilty" item in the null hypothesis.)

a. Your *p*-value suggests that the machine is okay, so you fail to reject Ho. But the machine really isn't okay, so Ho turns out to be false. Failing to reject Ho when Ho is false is a Type II error. You missed a chance to detect a problem.

TIP

A Type II error is a missed detection, or a missed opportunity, which is much easier to remember than "you fail to reject Ho when Ho is false."

b. The impact of making a Type II error in this situation is that you let the machine keep making widgets that don't meet specifications. You miss a detection of a problem, which costs people money and time in the long run.

4 Statistical Studies and the Hunt for a Meaningful Relationship

IN THIS PART . . .

See how researchers design and conduct various studies.

Practice spotting a bad study right off the bat.

Explore relationships between two categorical variables with two-way tables and two quantitative variables with correlation and regression.

» Evaluating the selected sample

» Spotting problems in the data collection process

» Examining survey results the right way

Chapter 15

Examining Polls and Surveys

Polls and surveys are a part of our daily lives. Pollsters ask you over the phone or through the mail to participate in a survey, or you read or hear about the results of the latest poll (from fashion to news, you can find polls on almost every topic). Surveys are very powerful, both in good and bad ways, and being able to dissect a survey from beginning to end helps you decide whether a survey is worthy of your participation or whether the results you see and hear are believable.

You can break the survey process down into a series of ten steps:

1. **State the purpose of the survey.**
2. **Define the target population (the group you intend to make conclusions about).**
3. **Choose the type of survey you want to administer (mail, telephone, interview, and so on).**
4. **Design the questions you plan to ask the survey participants.**
5. **Consider the timing of the survey.**
6. **Select the sample of individuals who will participate in the survey.**
7. **Collect the data from the participants.**
8. **Follow up with people who don't respond.**
9. **Organize and analyze the data according to the purpose of the survey.**
10. **Interpret the results and draw conclusions.**

I focus on each of these steps in this chapter through scenarios that help you practice deciphering and discerning the good from the bad in terms of what goes into surveys and what comes out of them.

Planning and Designing a Survey

You should take all the steps in the introductory list into account during the plan and design phase of any survey. If you don't think about how you plan to analyze your data before you collect it, you may be in for a big problem come analyzing time. In this section, I cover Steps 1 through 5, the planning and design phase of a survey. See the following for an example of looking at how the steps come into play in a real survey situation.

EXAMPLE

Q. Why is it important to state the purpose of a survey before you conduct it?

A. I can think of two reasons. First, if you state the purpose of the survey, you give your audience a clear idea of what you want to determine. Second, stating the purpose of the survey helps keep the researchers on task, so they don't go off track finding information about people or topics that don't adhere to the original survey agenda.

1 You want to determine the extent to which employees use personal email in the workplace of a certain company. What's your target population?

2 Suppose that your target population is the homeless population in a certain city, and your purpose is to find out how they live from day to day. Explain what types of surveys are inappropriate and why, and suggest a reasonable format to get the information you need.

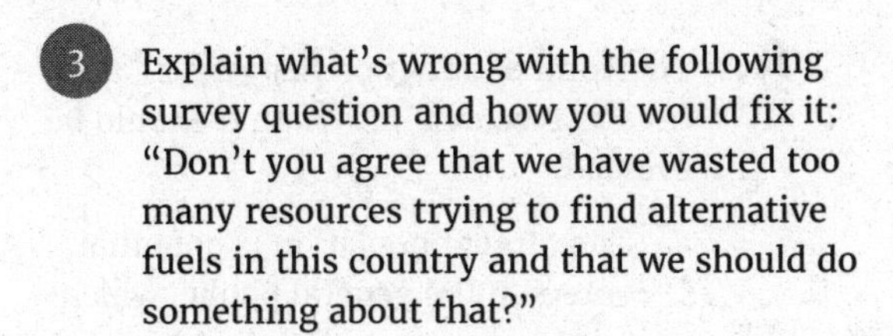

3 Explain what's wrong with the following survey question and how you would fix it: "Don't you agree that we have wasted too many resources trying to find alternative fuels in this country and that we should do something about that?"

4 What problems can you foresee if you conduct a survey to assess opinions on gun control right after a shooting takes place in a school?

Selecting a Random Sample

Selecting a sample can make or break the results of a survey, in terms of both bias (whether the results are on target or systematically off) and precision (how close the results are to the truth). Three criteria are important in selecting a sample that will give you the best results:

- Select a sample that represents the population you intend to study and make conclusions about (this is called the *target population*).
- Select a sample at random to avoid bias.
- Select a sample that's large enough to produce precise results.

To meet all these criteria, simply select a large *random sample.* A random sample is a sample chosen from the target population in such a way that every sample of the same size from the target population had an equal chance of being selected. A random sample avoids bias, should represent the target population, and, if large enough, yields precise statistical results (that is, results that are repeatable from sample to sample).

REMEMBER

The population you want to study and make conclusions about is called the *target population.* The population you end up sampling from is called the *sampled population.* You want these two populations to be the same, but unfortunately, the sampled population is often not the same as the target population, which causes bias. Before looking at the conclusions of a survey, take a look at who was actually studied, and see whether they represent the target population. This is one of those topics your instructor can really zoom in on.

See the following for an example of comparing the sampled population to the target population.

EXAMPLE

Q. Your psychology professor undertakes a study on the impact of negative political ads on potential voters in the general public. He finds participants for the study by offering his students extra credit.

a. What's his target population for the study?

b. Does this sample represent the target population? (What was the sampled population here?)

c. What's the impact of using this sample, in terms of the conclusions the prof may make?

A. This kind of situation happens all the time in research, although it should be avoided.

a. The target population is potential voters in the general public, which means anyone who's eligible to vote.

b. The sample doesn't represent the target population; it represents only a small part of it. (The sampled population is the students from this particular university who were enrolled in psychology at that time.)

c. Any results based on this sample of students should be made only about the students, not about the general public, or they will be biased and misrepresent the truth.

5 While you're surfing a news website, a pop-up asks you to participate in a survey to determine how people feel about reality television. What is the target population? What is the sampled population? Do they agree? Explain.

6 Bob wants to survey post-holiday shoppers, so he goes to the local mall, walks up to people at random, and asks them to participate in his survey. Is this a random sample?

7. Sue wants to conduct a telephone survey of people who live in her town. She opens the phone book to a random page and selects the first 100 names she sees. Is this a random sample?

8. A study, as described by the media, concludes that a new type of hair dye isn't harmful to your health. It turns out that it was based on only 14 people. Do you believe the results? Why or why not?

Carrying Out a Survey Properly

You've designed the survey and selected the participants. Now comes the process of implementing, or carrying out, the survey. This step presents its own set of challenges and problems to avoid.

See the following for an example of some of the problems that can come up during the survey process.

EXAMPLE

Q. Discuss two ways that survey participants can give incorrect information on a survey, and explain how you can minimize the problem in each case.

A. They can lie, or they can give the wrong answer by mistake (for example, they may be confused about a question). To minimize lying, make the survey anonymous (no one can link the people to their responses, not even you) or at least confidential (you promise not to divulge the connection between them and their answers). To minimize confusion, train the people asking the questions (phone survey or interviews) to be consistent, and make sure that mail surveys have clear questions and directions. Interviewers often have a script to ensure consistency between interviews.

9. Response bias occurs when survey participants give biased answers — either systematically higher than the truth or systematically lower than the truth. Give an example of two questions that could lead to these two types of response bias.

10. Suppose that you have two mail surveys; for the first, you mail out 10,000 surveys, and 1,000 people respond. For the second, you mail out 1,500 surveys and send reminder letters, and you ultimately receive 1,000 responses. Which survey do you think will produce more accurate results?

Interpreting and Evaluating Survey Results

The data analysis portion of the survey is the same as for any other data, and I discuss it in Chapters 2 through 4 and 9 through 11 of this workbook. Assuming that the data analysis has been done correctly and fulfills the purpose of the survey, the next step is to interpret the results and draw conclusions. Problems can creep in during this step, even if you deal with the most well-intentioned researchers.

Some of the more common errors made while drawing conclusions from surveys are

- Drawing conclusions about a population that's larger in scope than what the sample actually represents
- Claiming a difference exists between two groups when the difference isn't statistically significant
- Saying that "these results aren't scientific or statistically significant, but . . ." and then going on to present the results as if they hold scientific and statistical significance
- Trying to explain why the results are the way they are (for example, claiming a cause-and-effect relationship) without further study

See the following for an example of analyzing conclusions made from a survey.

EXAMPLE

Q. Suppose that a senator wants to introduce a bill to the U.S. Senate, and he uses the results of a survey posed to his constituents to help sell the bill to his colleagues. He introduces it by saying, "Americans really want this bill." Is this reasonable?

A. No, because he's generalizing beyond the sample. The survey represents only his constituents, not all American people. If he says, "My constituents really want this bill," that statement may be reasonable, assuming he took a large enough sample.

11 Suppose that poll results show 48 percent of voters favor Candidate A, and 52 percent favor Candidate B. Is it reasonable to say that Candidate B has a clear edge?

12 An evening television news program shows the results of its latest "call-in" poll, which explores how Americans feel about the job the president is doing. The station adds a disclaimer at the end of its piece, admitting that the results aren't scientific. Does this take care of the statistical problems?

13 Explain why a "call-in" poll to an evening news program constitutes an unscientific survey.

14 A survey asks college students a variety of questions about their daily routines and finds out that students who eat breakfast in the morning tend to have higher grades. The survey concludes that eating breakfast causes students to get better grades. Is this a proper conclusion?

Answers to Problems in Polls and Surveys

1. The target population is employees who have access to personal email in the workplace (not just employees who use personal email in the workplace). The target population is the group of individuals that you focus on making conclusions about. Because the extent of the use of personal email is the question, you must consider anyone who has a chance to send personal emails.

REMEMBER

Knowing your target population helps keep the results from being biased. Also, if the actual sample doesn't represent the target population, you know that the survey has a problem.

2. Because you're working with the homeless population, you need to realize that they have no permanent home, so they have no address or phone number for contact. A mail or telephone survey is out. Some type of personal interview with them is your best bet, in a nonintimidating setting. Some types of surveys are clearly inappropriate for certain situations, and you need to keep this in mind when you design a survey.

3. This is a leading question; you know exactly how the pollster wants you to answer it. The researcher clearly has an agenda, and the results of his/her survey are going to be biased. The way to avoid this is to word a poll question in a neutral fashion, such as the following: "What's your opinion on the issue of searching for alternative fuels — strongly agree, agree, no opinion, disagree, or strongly disagree?"

WARNING

Research shows that changing the wording of a survey question even slightly can greatly affect the results. Questions should be worded neutrally so as not to incur bias in the results.

4. Because the shooting just happened, people may have a stronger opinion about gun control than they had before the shootings, and research shows that, over time, after the tragedy fades from the mind, many of those folks go back to their original opinions. Although a survey directly after the incident may be informative and newsworthy, you have to put it into a larger timeline, keep the results in perspective, and be careful not to take advantage of certain situations.

REMEMBER

You have to take the timing of a survey under consideration, because it can greatly affect the results. Timing can mean what time of year the survey takes place or what time of day, for that matter. You shouldn't expect accurate results phoning office workers at home during the day, right?

5. The target and sampled populations do not agree. This sample represents only the people who visit the website and would be willing to participate in the survey. The target population is the people the survey implies to represent — all television watchers. The results of this web survey are biased because they don't represent the target population.

6. No. The definition of a random sample is a sample selected by giving every sample of the same size in the target population an equal chance to participate. The survey at the mall leaves out anyone who doesn't visit the mall that day. He also has no random mechanism for selecting people to participate, so he likely uses his own bias by approaching people who seem more friendly and willing, who walk slower, who don't look busy, and so on.

TIP

A random sample has a strict definition; to determine whether a sample is a truly random sample, check to see whether every sample of the same size in the target population had an equal chance of being selected.

7. No, she probably wouldn't choose the very first or very last page of the phone book to make her selection, so those pages don't have an equal chance of being selected. Second, choosing all the people from the same page is biased, because certain people may be related or come from a certain ethnic background if they share the same last name.

REMEMBER

A sample can't be "sort of" random or "close to" random. A sample is either random, or it isn't.

8. No. Because the study involves only 14 people, the population can't possibly be represented. The results would vary too much from sample to sample (which is what the margin of error measures).

TIP

To get a rough idea of the precision of a survey, take 1 divided by the square root of the sample size. For example, a survey of 400 people is precise to within $\frac{1}{\sqrt{400}} = \frac{1}{20} = 0.05$, or 5 percent. For only 14 people, the margin of error is roughly 26.7 percent!

9. Many questions are possible. Question 1: "How many fish did you catch on your fishing trip today?" This may entice people to overstate their actual results to avoid embarrassment or to impress people. Question 2: "How much money did you make last year, for income tax purposes?" This question can tempt people to understate the truth to avoid having to pay higher taxes.

10. Although both surveys received 1,000 responses, the second survey should probably have much more accurate results, because $1,000 \div 1,500$, or 67 percent, of the people surveyed responded, compared to only $1,000 \div 10,000$, or 10 percent, for the first survey. These percentages are called the *response rates*, and larger response rates typically mean more accurate results, because the survey leaves less chance for bias, and the sample is more representative of the target population.

REMEMBER

People who don't respond to a survey may be different from the people who do, which can bias the results. Always try for a high response rate, because a smaller survey with a high response rate is always better than a larger survey with a low response rate.

11. Not necessarily. You should expect the results to vary, and you need to take the margin of error into account. With a margin of error of 2 percent, Candidate A could have between 46 and 50 percent of the vote if you sample again, and Candidate B could have between 50 and 54 percent of the vote. Any larger margin of error causes overlap in the results; you can't say who wins because the percentages from the sample, although numerically different, aren't statistically different. (See Chapter 10 for more on margin of error.)

12. No. The results can still affect people; they may take them for the truth, because people regularly ignore disclaimers.

WARNING

Disclaimers about survey results not being scientific don't solve any statistical problems. They merely contaminate the pool of statistical information available. Use such survey results for entertainment purposes only.

(13) People who participate in a call-in poll don't represent the target population; they have to have a television, have it turned on, watch that program, and take the initiative to call in with an opinion. They also select themselves to participate in the survey (called a *self-selected sample* or a *volunteer sample*). Self-selected samples are biased, nonrandom samples.

(14) This conclusion cannot be made without further study that controls a host of other factors, such as exercise, sleep habits, dietary habits, study habits, IQ, and so on. This researcher goes too far in terms of trying to make an immediate cause-and-effect connection between certain findings. You can't make these types of conclusions without a study to look at the issue specifically. (See Chapters 16 and 17 for more on cause and effect.)

observational study

» Critiquing an experiment's manner of design

» Evaluating the results of an experiment correctly

Chapter 16
Evaluating Experiments

Experiments are one of the major vehicles for data collection. To assess whether the results of a hypothesis test, confidence interval, or regression analysis are credible and believable, you must first examine the collection of the data. I have a saying: "Garbage in equals garbage out." It means that no matter how sophisticated your graphs, charts, and data analysis are, they mean nothing if you base them on biased and inaccurate data.

In this chapter, you determine whether an experiment is well done and the impact the process has on the results.

Distinguishing Experiments from Observational Studies

Observational studies and experiments are vastly different in terms of the work that goes into them (their design and execution) and what comes out of them (the conclusions you can make). You need to be able to distinguish one type of study from another.

An *observational study* is just as it sounds: a study where the researcher merely observes the subjects and records the necessary information. The researcher doesn't intervene, impose treatments, or place restrictions or controls on the subjects. An *experiment* is the opposite situation. An experiment deliberately imposes some sort of treatment or condition on the subjects, controls the environment for other factors that may otherwise influence the results, records how the subjects respond to the treatments, and compares the results.

Because designed experiments are so controlled and restricted, you can say more in terms of your conclusions than after an observational study, where many other factors have a chance to confuse or confound the results. Designed experiments are better equipped to detect true cause-and-effect relationships (a change in one variable causes a change in the other variable); observational studies can say only that a relationship exists and that it may or may not be causal.

See the following for an example of distinguishing an experiment from an observational study.

Q. A study follows two groups for one year. The first group routinely uses antibacterial soap, and the other group routinely uses regular hand soap. The researchers record and compare the number and severity of illnesses for the two groups. Is this an observational study or an experiment?

A. This is an observational study, because the researchers didn't ask the people to use a certain type of soap; the groups used personal preference. The study didn't impose treatments or control other factors.

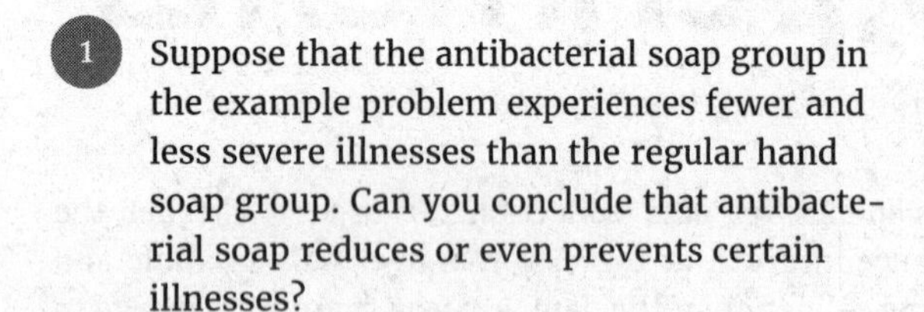

1. Suppose that the antibacterial soap group in the example problem experiences fewer and less severe illnesses than the regular hand soap group. Can you conclude that antibacterial soap reduces or even prevents certain illnesses?

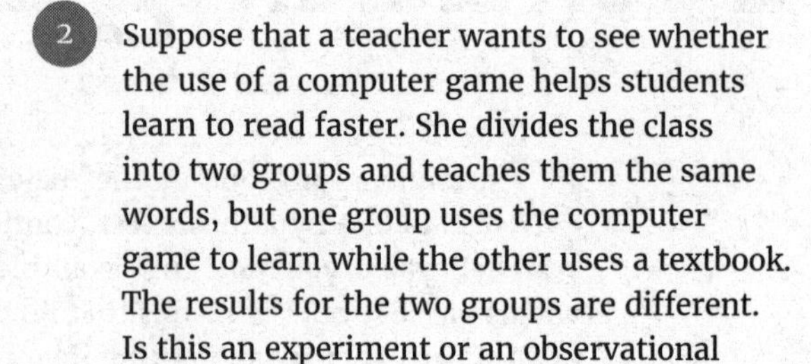

2. Suppose that a teacher wants to see whether the use of a computer game helps students learn to read faster. She divides the class into two groups and teaches them the same words, but one group uses the computer game to learn while the other uses a textbook. The results for the two groups are different. Is this an experiment or an observational study?

3. Suppose that you deem an experiment unethical; for example, the experiment forces some people to smoke two packs of cigarettes a day to see whether they develop lung cancer. What should you do to show this cause-and-effect relationship in humans?

4. Suppose that a group of people who normally take vitamin C each day experience fewer colds than a group of people who normally don't take vitamin C each day.

   a. What type of study is this?

   b. What other variables may explain the difference in the number of colds for the two groups?

Designing a Good Experiment

Every good experiment has three main features:

- The subjects are randomly assigned to treatments to avoid bias.
- The experiment controls as many confounding variables as possible. (A confounding variable is an unstudied or ignored variable that can influence the results.)
- The experiment has to take place enough times to ensure that the results have a high level of precision.

Data from a good experiment also has three qualities:

- **Validity:** It makes sense to collect this kind of data to answer the question posed.
- **Unbiasedness:** The data collected isn't systematically higher or lower than the truth.
- **Reliability:** The results are more likely to be repeatable if you conduct the experiment all over again (because the sample size is large enough to produce the necessary precision).

REMEMBER

An experiment's design and execution can make or break it in terms of the results you get and the conclusions you can reach. Putting all these ideas together into one list, you can evaluate an experiment based on whether it follows the criteria. A good experiment

- Includes a large enough sample size to produce precise results
- Chooses subjects that most accurately represent the target population
- Assigns subjects randomly to the treatment group(s) and the control group
- Controls for possible confounding variables
- *Double-blinds* the study to avoid bias (Double-blind means neither the subject nor the researcher knows who got what treatment.)
- Collects fair and accurate data (valid, reliable, and unbiased data)
- Contains the proper data analysis
- Has conclusions that don't go beyond the scope and limitations of the experiment

The first six criteria fall under the heading of the design of the experiment; I address them in this section.

See the following for an example of critiquing the results of an experiment.

EXAMPLE

Q. Suppose that an experiment shows that a new weight-loss plan works for seven out of ten people who try it. Because it works for 70 percent of the people who try it, are the results meaningful?

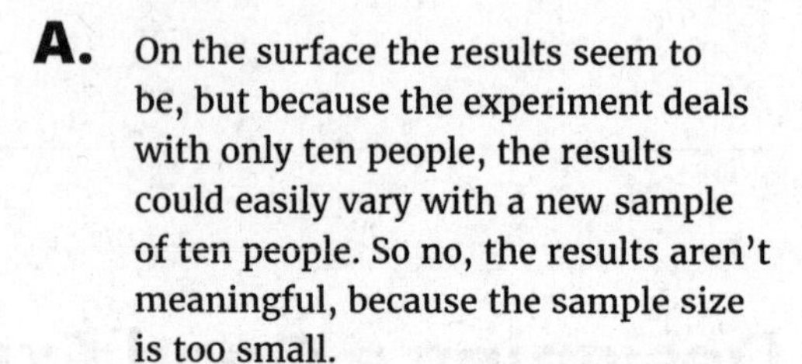

A. On the surface the results seem to be, but because the experiment deals with only ten people, the results could easily vary with a new sample of ten people. So no, the results aren't meaningful, because the sample size is too small.

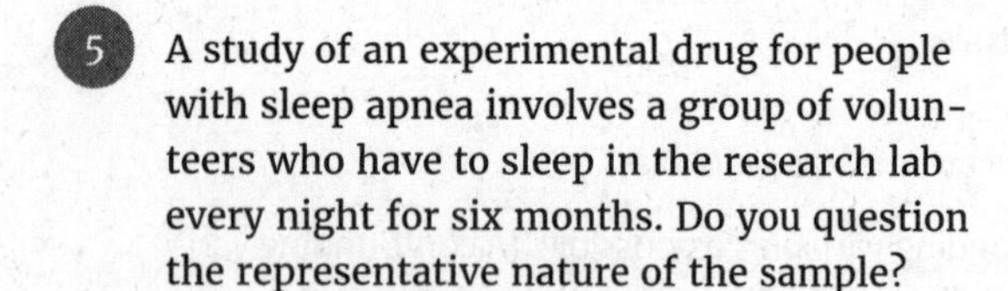

5 A study of an experimental drug for people with sleep apnea involves a group of volunteers who have to sleep in the research lab every night for six months. Do you question the representative nature of the sample?

6 A teacher wants to compare two methods of teaching math, one using technology and the other using a traditional approach. She needs to teach 25 of the 50 students with the technology method, and she asks for volunteers. Any problems with the design of the experiment?

7 An experiment compares two weight-loss programs (Plans 1 and 2) for overweight people. Of the 100 volunteers, researchers randomly assign half to each program. They record weight changes after six weeks but collect no other data. Plan 1 people lose 25 percent more weight than participants on Plan 2, so researchers deem Plan 1 the better plan. Name two confounding variables and explain why they are confounding.

8 A researcher is so excited about his experiment that compares blood pressure of dogs on a new drug to a control group that he watches and records all the information he can on the drug group and ignores the control group. What did he do wrong, and what's the impact of this error?

9 A study examines the effects of a new type of medicine on mild headache sufferers. The control group receives no drug, and the other participants take the new medicine. People on the medicine say they feel better right away. Any problems with this experiment?

10 A scale used in a weight-loss experiment is 5 pounds off in the positive direction. What kind of data problem does this pose, and what impact does the inaccuracy have on the results?

11 And for further proof that the researchers need a new scale, they find that it gives different results when the same person gets off and immediately gets back on. What kind of data problem does this pose, and what impact does this have on the results?

12 A teacher comparing the reading ability of his students using a computer program versus a traditional approach assesses and compares their progress by looking at their grades in gym class. What kind of data problem does he create here, and what impact does this have on the results?

Looking for Cause and Effect: Interpreting Experiment Results

Sound experiments have appropriate data analyses for the collected data and answer the posed question. Choosing the right analysis for the data is something you examine in Chapters 6 through 8 and 11 through 13. After you analyze the data, you can make conclusions. In any study, you don't want to overstate the results and mistakenly apply them to a larger population than what the sample actually represents. Another common mistake analyzers make is to jump to conclusions about cause and effect prematurely, explaining why they get the results without sufficient data. See the following for an example of correctly interpreting the results of an experiment.

EXAMPLE

Q. A study shows that people who achieve PhDs are less likely to develop memory loss diseases such as Alzheimer's than professionals without a PhD. Does this mean that if you want to avoid getting Alzheimer's, you should get your PhD?

A. Not necessarily. Although this study does show that a relationship exists, it doesn't automatically imply cause and effect. This study obviously couldn't have been an experiment, because it isn't ethical to force people to get PhDs or force them not to, so the possible confounding variables are many. You can appropriately state that a link exists and that researchers need to study it further. Perhaps you can formulate an experiment that mimics this dynamic.

13 An experiment concludes that eating an egg a day doesn't raise your cholesterol (as researchers once suspected). The experiment involves healthy, young males on low-fat diets. Explain what's wrong with the conclusions of this study.

An experiment shows that rats that receive a big piece of cheese at the end of a maze get to the cheese faster than rats that receive a small piece of cheese. Does this mean you can motivate rats to learn faster, like humans?

Answers to Problems in Evaluating Experiments

1. The words "prevent" and "reduce" imply a cause-and-effect relationship, which can't be determined through an observational study. Many other variables could account for the difference in results, such as health and hygiene habits, frequency of hand washing, temperature of the water used, length of hand washing, medical history, stress level, and so on.

2. The teacher runs an experiment because she assigns students to a treatment (computer game method or textbook method) and compares the results. The teacher controls the content. (Now, whether she ran a well-designed experiment is another issue.)

3. In certain cases, conducting an experiment is unethical. Although an observational study isn't the optimal study to conduct, it may be your only choice. And the evidence for a cause-and-effect relationship (although not directly proven) does mount as researchers conduct more and more observational studies, trying to explore the relationship between smoking and the occurrence of lung cancer by looking at many different groups of people in many different situations. In other words, a mountain of observational studies can add up to one well-designed experiment, if the need arises. (This is how researchers eventually showed that smoking causes lung cancer in humans; it wasn't so long ago.)

4. This study is an example of many studies that provide information for commercials; take these studies with a grain of salt.

   a. An observational study, because the people decide for themselves whether to take vitamin C. No treatment is imposed.

   b. Other uncontrolled variables that may influence the results include (but aren't limited to) health and hygiene habits, exercise, level of health consciousness, or diet.

REMEMBER

Other uncontrolled or ignored variables in a study that can affect the results are called *confounding variables.* Confounding variables are the reason observational studies aren't very strong. Experiments try to control variables as much as possible, so their results are more powerful.

5. This sample probably doesn't represent the entire population of people with sleep apnea. It includes only the subjects willing to sleep in a research lab for six months. Children, older people, and people who travel a great deal aren't fully represented, which can bias the results.

6. By asking for volunteers, the teacher puts bias in the results. Students who volunteer for the computer method are probably more comfortable with computers than students who don't. The assignments of students to a type of learning model have to be random.

7. The study has many possible confounding variables, including (but not limited to) the subjects' amount of weight to begin with, their amount of motivation, cost of the program, how well subjects adhered to the program, and medical issues. These items should be controlled for in the experiment, possibly by matching pairs of people and putting one in each program (randomly), along with controlling the outside environment. If they are not controlled for, these confounding variables can interfere with the amount of weight gained or lost, and researchers won't be able to tell whether the weight loss/gain is due to the program or to the confounding variables. Confounding variables confound or cloud the results, if you will.

8. The study isn't double-blind. Because the researcher knows who gets what treatment, his bias becomes part of the recorded data. A third party should assign dogs to treatments and not tell the researcher.

REMEMBER

An experiment is blind if the subjects don't know what treatment they receive and double-blind if the researchers also don't know who receives what treatment. Both blinding and double-blinding help reduce bias in the data-collection process.

9. Subjects often experience the *placebo effect.* That is, when subjects know they've taken a real (nonfake) treatment, the knowledge can affect their expectations and hence the data collected. Their expectations lead to biased results. To fix the problem, use a sugar pill for the control group and don't tell them which pill they receive.

REMEMBER

Researchers have a control group so they can determine the strength of the placebo effect and take that info into account for the real treatment group. This helps them determine the "real" effect of a real treatment. In cases where it might be unethical to use a fake treatment, a "standard treatment" can be used for comparison. An experiment is stopped in the case where the new treatment is found to be clearly better than the standard or placebo, so everyone suffering can get help, not just the treatment group.

10. The measurement instrument is biased, which leads to biased data.

11. The measurement instrument is unreliable, which leads to imprecise data.

12. The measurement is invalid, because grades in gym class don't measure reading ability. The invalidity leads to results that lack credibility.

13. The study is based on a sample that represents only young, healthy males on low-fat diets, but the conclusions say eggs don't raise *your* cholesterol (implying that the diet works for anybody). This mistake leads to severely biased results and misleading conclusions.

14. Maybe, but it could also mean that rats have a good sense of smell, and bigger bait smells stronger than smaller bait. The conclusion about motivation tries to answer the "why" question with insufficient data. Further studies must be done to address the "why" issue.

[illegible]

[illegible]

[illegible]

[illegible]

[illegible]

[illegible]

[illegible]

[illegible]

[illegible]

» Finding joint, marginal, and conditional probabilities

» Checking the independence of two categorical variables

Chapter 17

Looking for Links in Categorical Data: Two-Way Tables

In this chapter, you practice breaking down and analyzing the data presented in two-way tables. *Two-way tables* appear when two categorical variables are measured and you want to look at the relationship between the two variables. For example, how do males and females differ in terms of whom they voted for in the last presidential election?

You use appropriate probability notation to denote probabilities for two-way tables by applying letters to describe the groups comprising the categorical variables you want the probabilities for. For example, if set A stands for the group of females, A^c (the complement of set A, or everything that isn't in set A) stands for the group of males. Set B may represent people who vote for the incumbent president, and B^c represents people who don't vote for the incumbent president.

Understanding Two-Way Tables Inside and Out

Oftentimes, people measure different variables because they want to look for links or associations between them. If you want to find links, summarizing the individual variables isn't enough. The type of summary you use to look for links between two categorical variables is called a two-way table (also known as crosstabs). Two-way tables summarize the information from two categorical variables at once — such as gender and political party — so you can see the number of individuals in each combination of categories. For example, if you want to examine connections between gender (two categories: male and female) and political party (three categories: Democrat, Republican, and Independent), you can look at the percentage of Republican females, Republican males, Democrat females, Democrat males, and so on with a two-way table. The total number of possible combinations is $2 * 3 = 6$ in this case. Table 17-1 displays an example of a two-way table that shows political affiliation and gender for 200 people.

Table 17-1 The Gender and Political Affiliation of 200 Survey Participants

Gender	Democrat	Republican	Independent	Totals (Gender)
Male	35	55	10	100
Female	55	35	10	100
Totals (Political Affiliation)	90	90	20	200 (grand total)

The first step to success with two-way tables is to understand their organization and how you can get information from them. Each combination of categories is called a *cell*. Table 17-1 contains six cells: male Republicans, male Democrats, male Independents, female Republicans, female Democrats, and female Independents. You can see how many people are in each cell by intersecting the row and column of the groups that you want to look at and finding the number of people in that combination. For example, to see how many male Republicans are in the survey, intersect the row for males with the column for Republicans to find the number 55. Similarly, you see 35 male Democrats and 10 male Independents. For the females, you see 55 female Democrats, 35 female Republicans, and 10 female Independents.

You can also look at the *marginal totals,* which represent the total number in each row or column separately. You find marginal totals in the margins of the tables along with a heading called "Total." For example, in Table 17-1, the marginal total for males is 100; the marginal total for Democrats is 90. The *grand total* is the total of all the cells in the table (the total sample size). The grand total sits in the lower right-hand corner of the two-way table. In this example, the grand total is 200.

TIP

For any two-way table problem, you should write in the marginal totals and the grand total first thing. Then you can answer almost any question that comes up.

See the following for an example of depicting the individual cells in a two-way table.

EXAMPLE

Q. Suppose that a researcher divides a sample of 100 cars into groups according to their number of bumper stickers (three or fewer versus more than three) and the ages of the cars (5 years old or fewer versus more than 5 years old) and summarizes the results in the following table.

Number of Bumper Stickers	Age of Car $\leq$ 5 Years	Age of Car > 5 Years
0 – 3 bumper stickers	30	15
> 3 bumper stickers	20	35

a. Describe each cell in the two-way table and the number it contains.

b. Write in the marginal totals and the grand total and interpret them.

A. Here's how all the bumper-sticker info breaks down:

a. Thirty cars in the sample have few bumper stickers and are newer cars. Fifteen cars have few bumper stickers and are older cars. Twenty cars have a lot of bumper stickers and are newer cars. Thirty-five cars have a lot of bumper stickers and are older cars.

b. The marginal totals are included in the following table. The total number of cars in the sample with fewer bumper stickers (fewer than 3) is 45 $(30+15)$; the total cars with a lot of bumper stickers (3 or more) is 55 $(20+35)$. The total number of newer cars (less than 5 years old) is 50 $(30+20)$; the total number of older cars (5 years old or more) is 50 $(15+35)$. The grand total of all cars in the sample is 100.

Number of Bumper Stickers	Car $\leq$ 5 Years	Car > 5 Years	Totals (Stickers)
0 – 3 bumper stickers	30	15	45
> 3 bumper stickers	20	35	55
Totals (age)	50	50	100
			(grand total)

1 A medical researcher measures the dominant hand against gender for a group of 42 toddlers and shows the results in the following table.

Gender	Left-Handed	Right-Handed
Males	4	24
Females	2	12

a. How many toddlers are male?

b. How many toddlers are right-handed?

c. How many toddlers are right-handed and male?

2 For the gender and political affiliation data summarized in Table 17-1, find the marginal totals for both variables and interpret.

3 Find the marginal totals for the toddler data table in Question 1 and interpret.

It's important to keep the big picture of a two-way table in mind and not get lost in the individual cell-by-cell breakdowns.

a. If you sum all the marginal totals for the rows of a two-way table, what should you get?

b. If you sum all the marginal totals for the columns of a two-way table, what should you get?

Working with Intersection, Unions, and the Addition Rule

The *joint probability* is the probability that one selected individual falls into a particular row and column combination — for example, the chance that a house has both a two-car garage and a jet bathtub. To find a joint probability for a particular cell, you take the *cell count* (the number of individuals in that particular cell) divided by the grand total. If you sample 100 homes and 15 have both a two-car garage and a jet bathtub, the joint probability is $15 \div 100 = 0.15$ percent. Of all the homes you sample, 15 percent have both a two-car garage and a jet bathtub. The notation for the joint probability of A and B is $P(A \cap B)$. *Note:* The joint probability is often just written as P(A and B); I use both types of notation in this chapter so you can get used to both.

TIP

Joint probabilities always use the grand total in the denominator because you select someone or something from the entire group. On the flip side, they always use the cell count in the numerator because the individuals or items jointly fall into the two categories of interest (yes, the houses have two-car garages, and, yes, they have jet tubs). When you see the word *and* or *both* in a problem, you usually work with joint probability.

The *union probability,* $P(A \cup B)$, is the probability that either A or B or both occur. You denote this typically as P(A or B). For example, suppose that you roll a six-sided die. Let A be the event where you get an odd number (1, 3, or 5), and let B be the event where you get a number less than 3 (so 1 or 2 would qualify there). The probability of A or B is the event that you either get an odd number, a number less than 3, or both. That means you could have rolled a 1, 2, 3, or 5, and that probability is 4 out of 6, or 0.67. Notice that "1" appears in both events, but you count it just once when finding the probability of A or B.

REMEMBER

The word *or* in everyday language means "one or the other," but the word *or* in statistics means "either, or, or both."

The *addition rule* is a big one in probability. It shows you how to find the probability of A or B (in other words, the union probability). The addition rule says $P(A \cup B) = P(A) + P(B) - P(A \cap B)$.

Notice that you subtract the intersection probability in the addition rule. That's because the intersection was being counted twice, once as a part of P(A) and once as a part of P(B). You don't want it to be counted twice, so you have to subtract it out one time. For example, in Table 17-1, if you want the probability that someone is a male or a Democrat, you may think that you should take $P(M) + P(D)$ and be done with it. But there are some male Democrats out there (35 of them, to be exact, in this study), and they were each counted twice: once as a male and again as a Democrat. So you need to subtract them out once to balance things back out. In the end, you have P(male or Democrat) = $(100 \div 200) + (90 \div 200) - (35 \div 200) = 0.50 + 0.45 - 0.175 = 0.775$ or 77.5 percent.

A special case occurs when two events have no intersection. When $P(A \cap B) = 0$, you say A and B are mutually exclusive. In the case where you have mutually exclusive events, the probability of A or B is easier to find. It's just $P(A) + P(B)$ because $P(A \cup B) = P(A) + P(B) - 0$.

TIP

If you're asked to find a probability, most instructors don't mind you leaving the answer either as a percent (between 0 and 100) or as a proportion (decimal value between 0 and 1). Technically, however, probability is a number between 0 and 1 — the same as a proportion, which you're supposed to leave in decimal form. Ask your instructor in which format she wants you to have your answer.

The following is an example involving unions and intersections.

EXAMPLE

Q. Using the data from Table 17-1 and appropriate probability notation, identify and calculate the following:

a. What's the probability that an individual is both a Republican and a female?

b. What's the probability that an individual is Republican or female?

c. What proportion of the participants are female Independents?

d. What percentage of the participants are female Democrats?

e. Let D be Democrat and M be male. What's the probability of $P(D \cap M)$?

A. Question wording is subtle but important here.

a. The probability of being both a Republican (R) and a female (F) is denoted $P(R \cap F)$ and is the cell count for the Republican-female combination (35) divided by the grand total (200), which gives you $P(R \cap F) = 35 \div 200 = 0.175$, or 17.5 percent.

b. The probability of being a Republican or female is a union probability, so you use the addition rule. This probability is $P(R) + P(F) - P(R \text{ and } F) = (90 \div 200) + (100 \div 200) - (35 \div 200) = 0.45 + 0.50 - 0.175 = 0.775$ (or 77.5 percent).

c. The proportion of the participants who are female (F) Independents (I) means the fraction of all the female Independent participants. Again, you take the cell count divided by the grand total, so $P(F \cap I) = 10 \div 200 = 0.05$.

d. The percentage of the participants who are Democrat (D) and female (F) is a joint probability. You take the cell count for Democrat-female (55) and divide by the grand total (200) to get $P(D \cap F) = 55 \div 200 = 0.275$, or 27.5 percent.

e. This is asking for the joint probability of being Democrat (D) and male (M) at the same time, which is $P(D \cap M)$. This probability is $35 \div 200 = 0.175$.

5 Using the car bumper sticker data (see the previous section's example problem) and appropriate probability notation, identify and calculate the following:

a. What percentage of the cars are new and have a lot of bumper stickers? (Think mountain traveler and adventurer.)

b. What percentage of the cars have a lot of bumper stickers and are old? (Think 1960s vans.)

c. What percentage of the cars have a lot of bumper stickers or are old?

d. What percentage of the cars are old with a lot of bumper stickers?

6 Using the toddler data table in Question 1 and appropriate probability notation, identify and calculate the following:

a. What percentage of the toddlers are right-handed males?

b. What percentage of the toddlers are right-handed females?

c. Suppose that you want to see whether you can find a relation between gender and dominant hand. Can you compare your answers to parts a and b to come to a conclusion?

d. Find two events in this table that are mutually exclusive.

7 Using the toddler data table in Question 1, describe the toddlers in the sample with joint probabilities only.

8 If you find the joint probabilities for each of the cells in a two-way table, what should they sum to?

Figuring Marginal Probabilities

The *marginal probability* is the probability that one selected individual falls into one particular row or one particular column. Marginal probability looks at only one of the variables in the table — for example, the chance that a house has a two-car garage or the chance that a house has a jet bathtub. To find a marginal probability for a particular value of a row variable, take the *row count* (the number of individuals in that particular row) divided by the grand total. Similarly, to find a marginal probability for a particular column variable, take the column count (the marginal total for that column) divided by the grand total.

If you sample 100 homes and 35 have a two-car garage and 25 have a jet tub, the marginal probability for a two-car garage is $35 \div 100 = 0.35$, or 35 percent, and the marginal probability for a jet tub is $25 \div 100 = 0.25$, or 25 percent. Of all the homes you sample, 35 percent have a two-car garage and 25 percent have a jet tub. ***Note:*** Some of the homes may have both, but that involves a joint probability. Marginal probabilities look at one variable at a time. The notation for the marginal probability of A is P(A).

TIP

Marginal probabilities always use the grand total in the denominator, because you select someone or something from an entire group. On the flip side, they always use either the marginal row total or the marginal column total in the numerator, because you want the total number of individuals or items in the sample with the one characteristic of interest. When the question asks about only one of the variables, you usually work with marginal probability.

REMEMBER

You can interpret P(A) in different ways. It means the proportion of the entire group that belong to Group A, and it also means the chance of choosing one person or item from the entire group and identifying the person or item as a Group A member. Proportions and probabilities apply to groups and to individuals.

See the following for an example of calculating marginal probabilities.

EXAMPLE

Q. Using the data from Table 17-1 and the appropriate probability notation (for example, unions, intersections, and conditional probabilities), identify and calculate the following:

a. What's the probability that a participant is Republican?

b. What proportion of the sample are the females?

c. What percentage of the sample are Democrats?

A. Marginal probabilities are found by taking the number in the group of interest divided by the grand total (the total sample size).

a. The probability that a participant is Republican, P(R), is equal to the total number of Republicans (a marginal total, 90) divided by the grand total (200): $P(R) = 90 \div 200 = 0.45$, or 45 percent.

b. The proportion of females is the same as the probability that a participant is a female, P(F). You take the total number of females (a marginal total, 100) divided by the grand total (200): $P(F) = 100 \div 200 = 0.50$, or 50 percent.

c. You're looking for P(D), which is the number of Democrats (90) divided by the grand total (200), which gives you $90 \div 200 = 0.45$, or 45 percent.

9 Using the car data from the bumper sticker table earlier in this chapter and the appropriate probability notation, identify and calculate the following:

a. What percentage of the cars are newer cars?

b. What's the proportion of older cars?

c. What percentage of the cars have a lot of bumper stickers?

d. What's the probability that a car doesn't have a lot of bumper stickers?

10 Using the toddler data from the table in Question 1 and the appropriate probability notation, identify and calculate the following:

a. What percentage of the toddlers are right-handed?

b. What percentage of the toddlers are female?

c. What proportion of the toddlers are left-handed?

d. What's the chance of finding a male toddler from the sample?

11 Using the same toddler data, describe the toddlers with marginal probabilities only.

12 Which marginal probabilities that you find in a two-way table should sum to 1?

Nailing Down Conditional Probabilities and the Multiplication Rule

Conditional probabilities break a group into smaller subgroups and find probabilities within the subgroups. You use them for comparison purposes; for example, suppose that 60 percent of voters voted for the incumbent president in the last election. If you break it down by gender, you may find that 80 percent of women voted for the incumbent versus only 20 percent of the men. When you break a population into subgroups, probabilities can change.

For example, suppose that you know a house has a two-car garage, and you want to know the chance that it also has a hot tub. You aren't looking at all homes anymore, only those that have a two-car garage. It may be safe to assume that if a home has a two-car garage, the owners may be more likely to own a hot tub than owners of a home without a two-car garage. To find a conditional probability, you take the *cell count* (the number of individuals in a particular cell) divided by the marginal total for the group the individuals are already in. Suppose that you sample 50 homes that have a two-car garage, and 15 of them have a hot tub. The conditional probability is $15 \div 50 = 0.30$, or 30 percent. Of all the two-car garage homes you sample, 30 percent of those with a two-car garage have a hot tub.

TIP

Conditional probabilities always use marginal totals in the denominators because you select someone or something from a certain subgroup. You still use the cell count as the numerator because you still want the individuals or items that fall into the two categories of interest (yes, the homes have two-car garages, and, yes, they have hot tubs). When you see words like *given*, *of*, or *knowing* in a problem, you usually work with a conditional probability.

The notation you use for the conditional probability of A, given the individuals or items are in group B, is $P(A \mid B)$. Conversely, if you know the individuals or items are in group A and you want the probability that they also fall into group B, you express the probability as $P(B \mid A)$. The order of letters in the notation for conditional probability is critical. ***Note:*** $P(A \mid B)$ doesn't mean P(A) divided by P(B). The "|" sign doesn't mean division; it means "given" or "conditioned on the fact that."

To find the conditional probability, you can use the following definition: $P(A \mid B) = \frac{P(A \cap B)}{P(B)}$. Notice that the denominator in the formula for $P(A \mid B)$ is P(B) as mentioned in the previous paragraph. The numerator is the intersection probability. For example, look at a deck of 52 cards. Each card in the deck is marked with one of four suits, one of two colors, and one of 13 face values. Suppose that you want to know the probability that a card is red, given that the card is a 2. That means you want $P(\text{Red} \mid \text{Two})$. Using the definition of conditional probability, you take $P(\text{Red} \mid \text{Two}) = \frac{P(\text{Red} \cap \text{Two})}{P(\text{Two})} = \frac{\frac{2}{52}}{\frac{4}{52}} = \frac{2}{4} = 0.50$. Notice that this expression is the same as saying the deck has four 2s and two of them are red, so $P(\text{Red} \mid \text{Two}) = 2 \div 4 = 0.50$.

TIP

Some probability questions, such as the example in the previous paragraph, are easy to answer; others are very difficult. What's the difference? It all depends on two things: (1) what you're given in the problem to work with, and (2) what you're being asked to find. Sometimes you're asked to find something that's given to you or nearly given to you. Other times, the answer isn't that obvious. In these more complex situations, if you can identify what's being asked and what you're given in the problem in terms of probabilities, you have a good chance of solving it. Just look down your formula list (or in your memory, if a formula sheet isn't allowed) and choose a formula that contains both items. You can usually then plug in the items you're given and solve for what you need.

The *multiplication rule* is another big rule in probability. It shows you how to find the joint probability of A and B (when you can't figure it from scratch). The multiplication rule says $P(A \cap B) = P(A) * P(B \mid A)$. And because $A \cap B = B \cap A$, you can also say $P(A \cap B) = P(B) * P(A \mid B) = P(B \cap A)$. This rule is the same as the definition of conditional probability cross multiplied. It's the same formula, just a different format for a different type of question.

See the following for an example of identifying and calculating conditional probabilities.

EXAMPLE

Q. Using the data from Table 17-1 and the appropriate probability notation, identify and calculate the following:

a. What's the probability that a participant is Republican given that she is female?

b. What proportion of the females are Independents?

c. Let D = Democrat and F = female. Find and interpret $P(F \mid D)$ and $P(D \mid F)$.

A. I word these problems slightly differently, but they all require a conditional probability.

a. Here you choose an individual from the group of females, so you condition on the participant being female. You want to know the chance the female you select is a Republican, so the probability you want to find is $P(R \mid F)$, or the probability of being a Republican given that the individual is female. The denominator of this conditional probability is the total number of females (marginal row total, 100), and the numerator is the number of Republicans in that group (row, which is 35). $P(R \mid F) = 35 \div 100 = 0.35$, or 35 percent.

Note: You can also use the definition of conditional probability and say $P(R \mid F) = \frac{P(R \cap F)}{P(F)} = (35 \div 200)$ divided by $(100 \div 200) = 35 \div 100$ and notice that the 100s cancel out here (not a coincidence). It is easier in this case to do it just by using the numbers in the cells of the table right away. Each problem is different; remember to look at what you're given and what you're asked to find, and choose an approach that relates those two items somehow.

b. This problem states "of the females," which means you know you select only from the females. You want to know the chance that the female you select is an Independent, so you need to look for $P(I \mid F)$. The denominator is the number of females, 100 (marginal row total), and the numerator is the number of individuals in that row who are Independents, which is 10. $P(I \mid F) = 10 \div 100 = 0.10$.

c. P(F | D) means you select from among the Democrats in an effort to know the chance that the Democrat you select is female. The denominator is the total number of Democrats (marginal column total), 90, and the numerator is the number of individuals in this column who are females, 55. $P(F | D) = 55 \div 90 = 0.61$, or 61 percent. P(D | F) means the opposite; you select from among the females (of which you have 100), and you want the probability that the female you select is a Democrat. Of the females, 55 are Democrat. $P(D | F) = 55 \div 100 = 0.55$, or 55 percent.

Note: The denominator of a conditional probability is always smaller than the grand total because you look at a particular subset of the entire group, and the subset is smaller than the grand total. However, be careful which total you work with. P(A | B) uses the total in group B as the denominator, and P(B|A) uses the total in group A as the denominator. P(A | B) isn't equal to P(B | A), in general.

13 Using the bumper sticker data from the first example problem in this chapter and appropriate probability notation, identify and calculate the following:

a. Let O = older cars and B = a lot of bumper stickers. Find and interpret P(B | O).

b. What percentage of the older cars have a lot of bumper stickers?

c. Of the older cars, what percentage have a lot of bumper stickers?

d. What's the probability that a car has a lot of bumper stickers, given its old age?

Using the toddler data from the table in Question 1 and appropriate probability notation, identify and calculate the following:

a. What percentage of the male toddlers are right-handed?

b. What percentage of the female toddlers are right-handed?

c. What percentage of the right-handed toddlers are male?

d. What percentage of the right-handed toddlers are female?

15 Using the toddler data from the table in Question 1 and appropriate probability notation, do the following:

a. Compare the right-handed female and male toddlers in the sample, using conditional probabilities (only).

b. Compare the right-handed toddlers in the sample, using conditional probabilities (only).

16 In a two-way table with variables A and B, does $P(A|B) + P(A|B^c) = 1$?

17 In a two-way table with variables A and B, does $P(A|B) + P(A^c|B) = 1$?

18 Explain why conditional probabilities allow you to compare two groups regarding a second variable where joint and marginal probabilities don't.

Inspecting the Independence of Categorical Variables

The reason researchers collect data on two variables is to explore possible relationships or connections between the variables. For example, if more women voted for the incumbent president in the last election than men, you find that gender and voting outcome are related. If you find a relationship between two variables, you say that they're dependent. If two variables have no relation, you say that they're independent.

Here is the definition of independence of two events A and B: Two events A and B are *independent* if $P(A \mid B) = P(A)$ or if $P(B \mid A) = P(B)$. That means knowing that one event happened does not influence the probability of the other event happening. The conditional probability and the marginal probability are the same.

If two events are independent, the multiplication rule becomes easier, too. If two events A and B are independent, $P(A \cap B) = P(A) * P(B)$.

These properties of independence can be used two ways: First, you can use the definition of independence to check for independence of A and B (if the condition does not hold, A and B are not independent). Second, if you know A and B are independent and you want the joint probability of A and B, you can use the multiplication rule (simplified) and just multiply the marginal probabilities together. Use the one that matches the information you're given in the problem.

TIP

It takes more work to show independence than to show dependence. If one of the situations you check doesn't hold, your work is done — the variables are dependent. If the first situation checks out, you have to keep checking the other conditions until you cover them all before you can say the variables are independent.

However, just because two variables are dependent doesn't mean they have a cause-and-effect relationship. For example, if you observe that people who live near power lines are more likely to visit the hospital in a year's time due to illness, it doesn't mean the power lines cause the illnesses. However, a well-designed experiment can show a cause-and-effect relationship (in cases where you can ethically conduct the experiment). What you can conclude about a relationship doesn't depend on the statistics that you calculate as much as on the design of the study and the method of data collection. Refer to Chapters 15 and 16 for more information on observational studies and experiments.

See the following for an example of checking for independence of two events.

EXAMPLE

Q. Suppose that 50 percent of the employees of a company say they would work at home if they could, and you want to see whether these results are independent of gender. You break it down and discover that 75 percent of the females and only 25 percent of the males say they would work at home if they could. Are gender and work preference independent? Describe your results by using statistical notation and the appropriate probabilities.

A. No, gender and work preference are dependent. Let H refer to the employees who want to work at home, and let F denote females. According to the information given, $P(H) = 50$ percent for the whole group, but when you look at only the females, the proportion who want to work at home rises to 75 percent, so $P(H|F) = 75$ percent. When you look at only males, the proportion who want to work at home lowers to 25 percent, so $P(H|M) = 25$ percent. Because $P(H)$ isn't equal to $P(H|F)$, gender and work preference are dependent (you don't even have to look at the males).

19 Using the data from Table 17-1, are gender and political party independent for this group?

20 Using the bumper sticker data in the first example problem in this chapter, are car age and number of bumper stickers related? (In other words, are these two events dependent?)

21 Using the data from the table in Question 1, does the dominant hand differ for male toddlers versus female toddlers? (In other words, are these two events dependent?)

22 Suppose that A and B are independent and P(A) = 0.6 and P(B) = 0.2. What is P(A and B)?

23 Suppose that you flip a fair coin two times, and the flips are independent.

a. What is the probability that you will get two heads in a row?

b. What is the probability that you will get exactly one head?

c. How does your answer to b change if the chance of a head is 0.75 (coin is not fair)?

24 Suppose that you roll a single die two times, and the trials are independent. What is the chance of rolling two 1s?

25 Suppose that A and B are independent and $P(A) = 0.3$ and $P(B) = 0.2$. Find $P(A \cup B)$.

26 Suppose that medical researchers collect data from an experiment comparing a new drug to an existing drug (call this the treatment variable), regarding whether it made patients' symptoms improve (call this the outcome variable). A check for independence shows that the outcome is related to the treatment the patients receive.

a. Are treatment and outcome independent or dependent in this case?

b. Do the results mean that the new medicine causes the symptoms to improve? Explain your answer.

27 Suppose that A and B are mutually exclusive. Does this mean A and B are independent?

28 Suppose that A and B are complements of each other. Does this mean A and B are mutually exclusive?

Answers to Problems in Two-Way Tables

1 To get the right numbers from a two-way table, identify which row and/or column you're in and work from there. But before you start, it's always a good idea to figure and write down the marginal totals and grand totals (see the following table).

Gender	Left-Handed	Right-Handed	Totals (Gender)
Males	4	24	28
Females	2	12	14
Totals (hand)	6	36	42 (grand total)

a. To get the total number of male toddlers, you sum the values across the "male" row to get $4+24=28$.

b. To get the total number of right-handed toddlers, you sum the values down the "right-handed" column to get $24+12=36$.

c. The number of right-handed male toddlers is the number of toddlers in the intersection of the "male" row and the "right-handed" column: 24.

2 Summing the values across Row 1 gives you the total number of males: $35+55+10=100$. Summing the values across Row 2 gives you the total number of females: $55+35+10=100$. Summing the values down Column 1 gives you the total number of Democrats: $35+55=90$. Summing the values down Column 2 gives you the total number of Republicans: $55+35=90$. Summing the values down Column 3 gives you the total number of Independents: $10+10=20$. *Interpretation:* This survey contains an equal number of men and women. Democrats and Republicans are equal in number, and each outweighs Independents (90 Democrats and 90 Republicans compared to 20 Independents).

3 Summing the values across Row 1 gives you the total number of male toddlers: $4+24=28$. Summing the values across Row 2 gives you the total number of female toddlers: $2+12=14$. Summing the values down Column 1 gives you the total number of left-handed toddlers: $4+2=6$. Summing the values down Column 2 gives you the total number of right-handed toddlers: $24+12=36$. *Interpretation:* This data set contains twice as many male toddlers as female toddlers. There are many more right-handed toddlers than left-handed toddlers (36 versus 6).

4 Marginal totals add up to the grand total, no matter which direction you sum them in (rows or columns).

a. If you sum all the marginal totals for the rows of a two-way table, you get the total sample size (otherwise known as the grand total). For example, in Question 3, the row totals are 28 (total male toddlers) and 14 (total female toddlers). These values sum to 42, or the total number of toddlers. The result makes sense because every toddler has to fall into one of the two categories in terms of his or her gender.

b. If you sum all the marginal totals for the columns of a two-way table, you also get the total sample size (the grand total). For example, in Question 3, the column totals are 6 (left-handed toddlers) and 36 (right-handed toddlers); the total size of 42 again makes sense because every toddler in the sample falls into one of these two categories in terms of his or her hand dominance. The table from Answer 1 shows what the marginal totals look like if you place them into the two-way table. (Where else would marginal totals be but out in the margins?)

TIP

Your first task when you look at a two-way table is to write down all the marginal totals in their proper places and make sure they add up to the grand total in the lower right-hand corner. The grand total is the total sample size.

(5) The word *and* is a good indicator that you're dealing with a joint probability. Also, look at what group you select the individuals from. "Of the cars" means you select them from the whole group, another indicator of a joint probability. Let B = cars with a lot of bumper stickers (so B^c represents cars without a lot of bumper stickers as the complement) and N = newer cars (so N^c represents older cars as the complement).

a. You want $P(N \cap B)$. The numerator is the cell count for the "newer, > 3 bumper stickers" cell (20), and the denominator is the grand total (100). The answer is $20 \div 100 = 0.20$, or 20 percent.

b. Now you want $P(B \cap N^c)$. The numerator is the cell count for the "older, > 3 bumper stickers" cell (35), and the denominator is the grand total (100). The answer is $35 \div 100 = 0.35$, or 35 percent.

c. This is asking for $P(B \cup N^c)$, and you use the addition rule to solve it. Take $P(B) + P(N^c) - P(B \text{ and } N^c) = (55 \div 100) + (50 \div 100) - (35 \div 100) = 0.55 + 0.50 - 0.35 = 0.70$, which is 70 percent.

d. Part d is the same as part b, but I word it a little differently. The answer is 0.35, or 35 percent.

TIP

One of the challenges of categorical variable problems is deciphering what the problem gives you and what you have to find, all in terms of probabilities. Practicing as many types of wordings of these problems and knowing what words to look for can really help.

(6) Let M = male, F = female, R = right-handed, and L = left-handed. (You can also use complement notation, but why confuse things?)

a. This part asks for a joint probability, even though it may appear that the toddlers come from the males to begin with. The key is the phrase "of the toddlers"; it means the individuals come from the whole group, so you're dealing with a joint probability. The probability you need to find is $P(M \cap R)$. Take the number of toddlers in the "male, right-handed" cell (24) divided by the grand total (42) to get $24 \div 42 = 0.57$, or 57 percent.

b. Here you want $P(F \cap R)$. Take the number of toddlers in the "female, right-handed" cell (12) divided by the grand total (42) to get $12 \div 42 = 0.29$, or 29 percent.

c. Comparing the results of parts a and b doesn't tell you whether gender and dominant hand are related. Even though approximately 60 percent of all toddlers are male right-handers and only 30 percent are female right-handers, that doesn't mean that more of the males are right-handed than the females. In this case, you find many more males in the sample than females, and using the total number of all toddlers in the denominator is unfair to the female group. To compare females to males, you split the toddlers into groups by gender and look at the percentage of right-handers in each group. For males, the percentage is $24 \div 28 = 85.7$ percent, and for females, the percentage is $12 \div 14 = 85.7$ percent. So the males and females have the same percentage of right-handers.

d. M and F are mutually exclusive events, because $P(F \cap M) = 0$. Similarly, L and R are mutually exclusive events.

REMEMBER

Don't use joint probabilities to check to see whether two variables are related in a two-way table. Split the groups and compare the percentages for each group. (These individual probabilities are called *conditional probabilities,* and I handle them in the section "Nailing Down Conditional Probabilities and the Multiplication Rule" in this chapter.)

7. When you use joint probabilities, your conclusions are somewhat limited because of the differences in the sizes of the subgroups. But in this example, you can say that $4 \div 42 = 9.5$ percent of the toddlers are male left-handers; $24 \div 42 = 57.1$ percent are male right-handers; $2 \div 42 = 4.8$ percent are female left-handers; and $12 \div 42 = 28.6$ percent are female right-handers.

8. Each of the cells should sum to 100 percent, because every individual in the sample belongs in exactly one cell, so the total has to add up to 100 percent. In the answer to Question 7, you can see that $9.5 + 57.1 + 4.8 + 28.6 = 100.00$ percent. (This is all subject to possible rounding error of course.)

9. Let B = cars with a lot of bumper stickers (so B^c represents cars without a lot of bumper stickers as the complement), and let N = newer cars (so N^c represents older cars as the complement).

a. The question asks for P(N). Take the marginal column total for the newer cars $(30 + 20 = 50)$ and divide by the grand total (100) to get 0.50, or 50 percent.

b. The percentage of older cars is 100 minus the percentage of newer cars, which is $100 - 50 = 50$ percent, because the groups are complements. (Or you can find the marginal column total for the older cars $[15 + 35 = 50]$ divided by the grand total [100], which is 0.50, or 50 percent.) Note, however, that the question asks for a proportion, so you should divide your percents by 100. In other words, the proportion of older cars is 0.50. Refer to Chapter 5 for more information on probability rules, complements, and so on.

REMEMBER

Some instructors are really picky about the proportion versus percentage thing, and others are not. Technically, they are different units to represent the same quantity. Suppose that you have 20 people out of 40. This can be written as 20 divided by 40 equals 0.50, which is a proportion (because proportions are decimals between 0 and 1). Or it can be written as 20 divided by $40 * 100$ percent, which is 50 percent (because percentages are always between 0 and 100). To be safe, ask your instructor how picky he is about this issue and which formats are acceptable for answers.

c. The question asks for P(B). Take the marginal row total for the cars with a lot of bumper stickers ($20+35=55$) and divide by the grand total (100) to get 0.55, or 55 percent.

d. You take $1-P(B)$, or $1-0.55=0.45$, or 45 percent, again by complements. Note that I word this problem differently in the sense that it asks for a probability, but its meaning is the same as the other problems in this section.

TIP

Be on the lookout for parts of problems that are really just complements of previous parts. Taking 1 minus a previous answer can save a lot of time. (This practice is a common one among teachers.)

(10) Let M = male, F = female, R = right-handed, and L = left-handed. (You can also use complement notation, but why confuse things?)

a. You want to find P(R). Take the marginal column total for the right-handers ($24+12=36$) and divide by the grand total (42) to get 0.857, or 85.7 percent.

b. You want to find P(F). Take the marginal row total for the females ($2+12=14$) and divide by the grand total (42) to get 0.333, or 33.3 percent.

c. You want to find P(L), which is the same as $P(R^c)$. You can take 1 minus the answer to part a to get $1-.857=0.143$.

d. You want to find P(M), which is the same as $P(F^c)$. You can take 1 minus the answer to part b to get $1-0.333=0.667$, or 66.7 percent.

(11) In the table in Question 1, you can say that the percentage of female toddlers is 33.3 percent, the percentage of males is 66.7 percent, the percentage of right-handers is 85.7 percent, and the percentage of left-handers is 14.3 percent.

REMEMBER

Marginal probabilities discuss only individual variables separately without examining the connection, so you have limited interpretation ability. Beware of people in the media reporting statistics from individual variables without examining the connection between them. If you don't examine all the cells of the two-way table, you miss a lot of information.

(12) The complements in a two-way table have probabilities that sum to 1 (as you see in Questions 9 and 10). Let A and A^c be the row values in the table (for example, males and females): $P(A)+P(A^c)=1$. Let B and B^c be the column values in the table (for example, right-handers and left-handers): $P(B)+P(B^c)=1$.

(13) Did you realize that all four parts of this problem are asking for the same thing, just with different wording? (Statistics is such an exact science, isn't it?) Let B = cars with a lot of bumper stickers and O = older cars for your notation. This problem is designed to help you get ready for possible wordings (and rewordings) of problems on your exams.

a. $P(B|O)$ means that given an older car, what's the chance of it having a lot of bumper stickers? The denominator of this probability is the total number of older cars (marginal column total, 50), and the numerator is the number of cars in the older-car column that have a lot of bumper stickers (35). So you have $P(B|O)=35\div50=0.70$, or 70 percent.

b. The word *of* tells you you've got a conditional probability and that you know the car is old, so that's what you're conditioning on (or putting in the back part of the formula). So again, you have $P(B|O)=35\div50=0.70$, or 70 percent. I can understand why someone would want to know this, don't you? Those old cars have that wild and windblown look about them . . .

c. Now the wording of this question sounds as if you're setting aside the older cars and examining their bumper stickers. In the end, you get the same answer as for the other parts of this problem, because that's exactly what conditional probability does — it sets aside the group. (I typically draw a circle around the row or column in a conditional probability to remind myself of that.) So again, the answer is $P(B|O) = 35 \div 50 = 0.70$, or 70 percent. Now, doesn't this makes sense to set older cars aside? Their bumper stickers are probably much more interesting, saying things like, "Honk if you love statistics!"

d. This wording is that warm and fuzzy old standby, using the word *given*. Older stat books still use this notation, but most of us have realized that people don't talk like that in the real world and have moved on to wordings like parts a–c in this problem. However, did I say that your statistics class was the real world? So again, you have $P(B|O) = 35 \div 50 = 0.70$, or 70 percent.

TIP

Notice that I change notation throughout this chapter to describe the same event at times. That's because as the focus changes in a problem, I like to change my notation. It's my notation after all, and it may as well be notation that I want. If the problem focuses on the fact that the cars are older and I want other probabilities relating to that, I use O to indicate older cars. If the problem is focusing on the newer cars for the most part except for a quick switch to old ones, I use N^c to indicate older cars. You should try the same; use whatever notation works and feels most comfortable for you — just be clear in defining it so your instructor knows what you're talking about.

(14) Let M = male, F = female, R = right-handed, and L = left-handed. (You can also use complement notation, but why confuse things?)

a. You want to find $P(R|M)$ because of the phrase "of the male toddlers," which means you select from among the males only (so M appears after the "|" sign in the probability). The denominator of this probability is the number of males (marginal row total, 28), and the numerator is the number of right-handed individuals in the row (24): $P(R|M) = 24 \div 28 = 0.857$, or 85.7 percent.

b. You want to find $P(R|F)$ because of the phrase "of the female toddlers," which means you select from the females only (so F appears after the "|" sign in the probability). The denominator of this probability is the number of females (marginal row total, 14), and the numerator is the number of right-handed individuals in the row (12): $P(R|F) = 12 \div 14 = 0.857$, or 85.7 percent. (The percentage of right-handers is the same for females and males — gender and dominant hand aren't related.)

Keep in mind that you can't use $P(R|M)$ to help you find $P(R|F)$ because the two groups (males and females) are independent, and conditional probabilities with different "denominators" don't add up.

c. Now you go the other way with your conditional probabilities (compared to part a) by dividing the toddlers into the left- and right-handed groups and selecting from there. In this part, you want to find $P(M|R)$ because of the phrase "of the right-handed toddlers," which means you select from the right-handers only (so R appears after the "|" sign in the probability). The denominator of this probability is the number of right-handers (marginal column total, 36), and the numerator is the number of male individuals in the column (24): $P(M|R) = 24 \div 36 = 0.67$, or 67 percent.

d. You want to find $P(F \mid R)$ because of the phrase "of the right-handed toddlers," which means you select from the right-handers only (so R appears after the "|" sign in the probability). The denominator of this probability is the number of right-handers (marginal column total, 36), and the numerator is the number of female individuals in the column (12): $P(F \mid R) = 12 \div 36 = 0.33$, or 33 percent.

Note: Your answers to parts c and d sum to 1 because c and d are complements. Knowing that you're in the right-hander group, you have to be either a male or a female, so you could take $1 - 0.67$ to get your answer for d.

REMEMBER

In general, $P(A \mid B)$ isn't equal to $P(B \mid A)$, as you can see by comparing your answers to parts a and c of this problem. Knowing which group is which and what the notation means is very important to solving these problems correctly.

15 This problem is asking you to do exactly what you did in Question 14. The trick is knowing that's what the question is asking you to do.

a. See the answers to Questions 14a and 14b.

b. See the answers to Questions 14c and 14d.

16 No. $P(A \mid B)$ and $P(A \mid B^c)$ aren't related because they condition on two different groups, like males and females. The two groups are independent, and you can't assume that they're related. $P(A \mid B)$ and $P(A \mid B^c)$ aren't complements because they don't condition on being in the same group. See the answers to Questions 14a and 14b for examples.

17 Yes, because both probabilities condition on the same group, and A and A^c are complements. The answers to Questions 14c and 14d show examples of events that are complements.

18 Marginal probabilities discuss only individual variables separately without examining the connection, so you have limited interpretation. For example, if you know that 50 percent of all people approve of a smoking ban and 20 percent of people are smokers, can you conclude that 50 percent of smokers approve of the smoking ban and 50 percent of smokers don't? No. All the smokers may oppose the ban, and all the nonsmokers may approve of it. Conditional probabilities break down the groups and compare them. You need conditional probabilities if you want to examine relationships among two categorical variables.

19 No. You can discover this in a couple of different ways. Suppose that you condition on gender and look at the percentage of Democrats, Republicans, and Independents. If gender and political party are independent, the percentages are the same for males and females, and the percentage equals the overall percentage of Democrats, Republicans, and Independents for the entire group. You know $P(D) = 90 \div 200 = 0.45$ and $P(D \mid F) = 55 \div 100 = 0.55$, so you know the two aren't independent. The percentages don't match up.

20 Yes, age and bumper stickers are related (or dependent). You can show their dependence in two different ways, depending on which variable you condition. First, the percentage of older cars, $P(O)$, is $50 \div 100 = 50$ percent. Of cars with a lot of bumper stickers, the percentage of old cars is $P(O \mid B) = 35 \div 55 = 64$ percent. More of the cars loaded with bumper stickers are older, so bumper stickers and car age are dependent. Another way you can show dependence is to look at the percentage of all cars with a lot of bumper stickers, $P(B)$, which is 55 percent, and compare it to the percentage of older cars with a lot of bumper stickers: $P(B \mid O) = 35 \div 50 = 70$ percent. More of the older cars are loaded with bumper stickers compared to the entire group.

TIP

To check to see whether A and B are independent, you can examine whether $P(A \mid B) = P(A)$, or you can examine whether $P(B \mid A) = P(B)$. Choose the situation that you feel comfortable working with. Most people prefer working by conditioning on the row variables rather than the column variables. It just seems easier.

21 Always watch for subtle changes in wording that mean exactly the same thing. Here I use a different kind of wording for the same type of independence problem. If dominant hand differs for the males versus the females, dominant hand is related to, or dependent upon, gender. So you need to determine whether gender and dominant hand are independent. Check to see whether $P(R \mid M)$ is equal to $P(R)$ or whether $P(R \mid F)$ is equal to $P(R)$. In this case, $P(R \mid M) = 24 \div 28 = 0.857$, and $P(R) = 36 \div 42 = 0.857$. Now you can see that $P(R \mid F) = 12 \div 14 = 0.857$, and $P(R) = 36 \div 42 = 0.857$. The percentage of right-handers for the male and female groups is the same as the percentage of right-handers in the entire group. Therefore, gender and dominant hand are independent for this group. That means there's no difference between the groups in terms of dominant hand. Gender and dominant hand are not related.

22 By the definition of independence of events A and B, $P(A \text{ and } B) = P(A) * P(B) = 0.6 * 0.2 = 0.12$.

23 Let H = heads and T = tails. Note that the probability of getting a head on a fair coin is 50 percent or 0.50, and the same with tails.

REMEMBER

Knowing that the probability of heads is 0.5 is important information that you had to figure out from the problem (it was not explicitly given to you). Instructors love to put these kinds of problems on exams, so watch for them and remember that if you aren't given the probability in the problem, it means you should be able to figure it out from the information given.

a. Here you want to get two heads. So what you're looking for is P(Head on 1st toss $\cap$ Head on 2nd toss = $P(HH)$. Because the flips are independent, you know that $P(HH) = P(H) * P(H) = 0.5 * 0.5 = 0.25$. This means that, in the long run, one-fourth of the time you'll see two heads when you flip a coin twice.

b. The probability of getting exactly one head in two tosses is P(HT or TH). By the addition rule, this equals $P(HT) + P(TH)$ because these events have no intersection. Now by independence of the two flips, you can say this equals $P(H) * P(T) + P(T) * P(H) = 0.5 * 0.5 + 0.5 * 0.5 = 0.25 + 0.25 = 0.50$. So 50 percent of the time you should expect exactly one head when you flip a coin twice.

WARNING

Resist the urge to say the answer is simply 0.5 because the probability of getting one head is 0.50. You haven't taken into account the fact that you had two tosses and what might have happened on that second toss. The sample space of this experiment is the four outcomes — HH, HT, TH, and TT — and you have to deal with both items in the pair, not just the one you're interested in.

c. The work to solve this part of problem is the same as for part b until you get to the last step, where you plug in the probabilities of heads and tails. There, you use $P(H) = 0.75$ and $P(T) = 1 - P(H) = 1 - 0.75 = 0.25$. You have P (exactly one head in two tosses of the unfair coin) is $0.75 * 0.25 + 0.25 * 0.75 = 0.1875 + 0.1875 = 0.375$.

24 This problem is similar to Question 23; you aren't explicitly given any probabilities in the problem, but don't let that get you stuck. Remember that the probability of getting a 1 on a fair die is 1 out of 6, so use $\frac{1}{6}$ as your probability. The question asks you to find the probability of rolling two 1s when you roll a fair die twice, so you need $P(1 \text{ and } 1) = P(1) * P(1)$ because the rolls are independent. Each $P(1)$ is $\frac{1}{6}$, so multiply them to get $(\frac{1}{6}) * \frac{1}{6} = \frac{1}{36} = 0.028$. Or you could have just realized that there are $6 * 6 = 36$ outcomes when you roll a die two times, and one of those outcomes is the (1, 1) outcome.

25 You're looking for P(A or B), which means you want $P(A \cup B) = P(A) + P(B) - P(A \cap B)$. This equals $0.3 + 0.2 - P(A \text{ and } B)$. You may think you're stuck here, but you aren't. A and B are independent, so P(A and B) is equal to $P(A) * P(B) = 0.3 * 0.2 = 0.06$. Substituting this in the problem, you get $0.3 + 0.2 - 0.06 = 0.44$ for the probability of A or B.

REMEMBER

You're usually given the information for a reason. In Question 25, you knew A and B were independent. Without realizing that means you can multiply P(A) times P(B) to get P(A and B), you would have been sunk. Don't let this happen to you on an exam. Have all those tools (the formulas, properties, definitions, and so on) ready to pull out when you need them. And be able to identify when you need them.

26 The manner of data collection directly affects what conclusions you can make.

a. Because the outcome is related to treatment, the two variables aren't independent; they are, therefore, dependent.

b. As long as the experiment is well designed (see Chapter 16), you can imply a cause-and-effect relationship. If this study is observational, or badly designed, the answer is no.

REMEMBER

The check for independence of two categorical variables is somewhat limited, in that you can really make conclusions only about the sample. You can't make broader conclusions to the general population without doing a hypothesis test. However, a hypothesis test for two proportions is equivalent to a hypothesis test for independence. So if you want to show that the variables have a statistically significant relationship (or not), do a hypothesis test for two proportions. See Chapter 13 for more information.

27 No. If A and B are mutually exclusive, their intersection is empty, so $P(A \cap B) = 0$. If A and B are independent, $P(A \cap B) = P(A) * P(B)$. This won't be 0 unless A or B is the empty set (a rare situation).

28 Yes. If A and B are complements of each other, their union is the entire sample space S, and their intersection is the empty set. That means they are mutually exclusive.

» Calculating and understanding correlation

» Finding the best fitting line (regression line)

» Forecasting with regression lines

» Testing the fit of the regression line

Chapter 18

Searching for Links in Quantitative Data: Correlation and Regression

In this chapter, you practice breaking down and analyzing data from two quantitative variables collected as pairs (*X*, *Y*). You also examine and quantify any linear relationship that may occur between the two variables. For example, can you use annual rainfall to predict corn production? What can you use to predict your total score in golf? And do fans consume more coffee at football games played in cold weather?

Relating X and Y with a Scatterplot

To make sense out of any type of data, you should organize it with a table, chart, or graph. In a case where you have data on two variables and both are quantitative (or numerical), such as height and weight, you organize it in a graph called a *scatterplot*. A scatterplot has two dimensions: a horizontal dimension (called the X-axis) and a vertical dimension (called the Y-axis). Both axes are numerical, and each contains a number line.

Plotting pairs of data on a scatterplot is similar to playing the game Battleship. Each observation has two coordinates; the first coordinate corresponds to the first piece of data in the pair (the X-coordinate; the amount that you go left or right), and the second coordinate corresponds to the second piece of data in the pair (the Y-coordinate; the amount that you go up or down). Intersect the two coordinates to pinpoint the spot where you place the point representing the pair of data for that observation.

You interpret a scatterplot by looking for linear trends in the data as you go from left to right. If the data resembles an uphill line as you move from left to right, this indicates a *positive linear relationship* — as X increases, Y increases. If the data resembles a downhill line as you move from left to right, this indicates a *negative linear relationship* — as X increases, Y decreases. If the data doesn't seem to resemble any kind of line, it has no linear relationship.

See the following for an example of making a scatterplot.

Q. Draw a scatterplot of the following small data set: (2, 3), (6, 8), (3, 5). Tell whether X and Y have a positive linear relationship.

A. See the following figure for a scatterplot of the data set. The data implies a positive linear relationship between X and Y, because as the X value increases, the Y value tends to increase along with it. (***Note:*** This small data set is for practice purposes only and wouldn't be large enough to make conclusions about X and Y.)

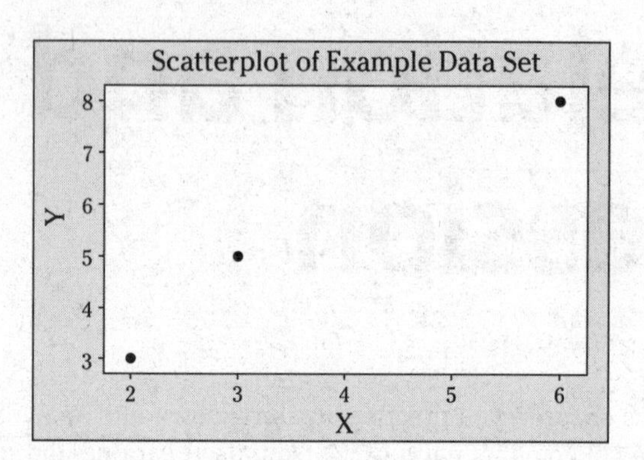

1 Describe the relationship between *X* and *Y* shown by the scatterplot in the following figure.

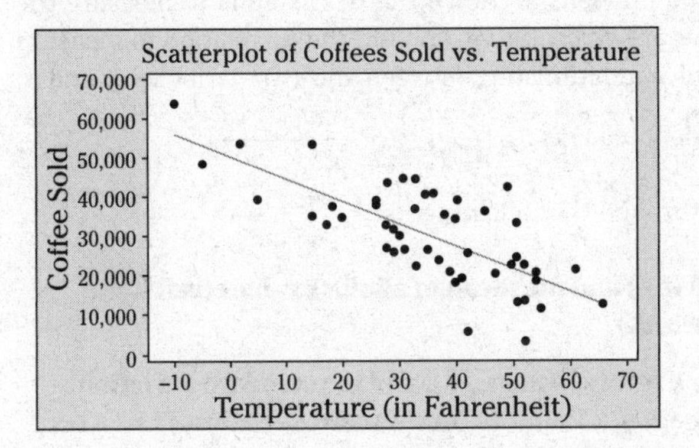

2 Describe the relationship between *X* and *Y* shown by the scatterplot in the following figure.

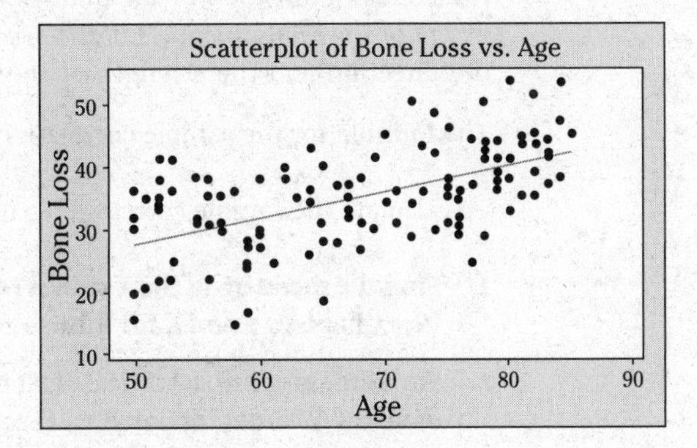

3 Describe the relationship between *X* and *Y* shown by the scatterplot in the following figure.

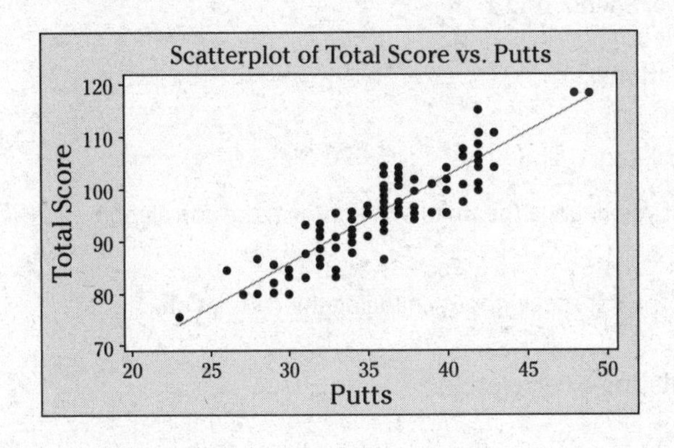

4 What does a scatterplot that shows no linear relationship between *X* and *Y* look like?

Toeing the Line of Correlation

After the organization of the data (see the previous section), the next step is to measure the extent of the relationship. If both variables are quantitative, you use the correlation to measure the direction and the strength of the linear relationship between the two variables X and Y. The formula for the sample correlation (denoted r) is $r = \frac{1}{n-1}\left(\frac{\sum(x-\bar{x})(y-\bar{y})}{s_x s_y}\right)$.

To calculate the sample correlation, run through the following list:

1. **Find the mean of all the *X* values (call it $\bar{x}$) and the mean of all the *Y* values (call it $\bar{y}$). (See Chapters 2 and 4 for more on means.)**
2. **Find the standard deviation of all the *X* values, denoted s_x, and the standard deviation of all the *Y* values, denoted s_y. (See Chapter 2 for more on standard deviation.)**
3. **For each (*x*, *y*) pair in the data set, take *x* minus $\bar{x}$ and *y* minus $\bar{y}$ and multiply them together.**
4. **Add all the products together to get a sum.**
5. **Divide the sum by s_x times s_y.**
6. **Divide the result by $n-1$, where *n* is the number of (*x*, *y*) pairs.**

Correlation is always between –1 and +1. If $r = -1$, you have a perfect downhill linear relationship between X and Y (as one goes up, the other goes down in a perfectly straight line). A correlation close to –1 indicates a strong downhill linear relationship. If the correlation is close to 0, the data has no linear relationship. If the correlation is close to +1, *X* and *Y* have a strong uphill linear relationship (as one goes up, so does the other). If the correlation is exactly +1, you have a perfect uphill linear relationship between *X* and *Y*.

TIP

Here are some useful properties of correlations:

- The correlation coefficient is always between –1 and +1.
- The correlation is a unitless measure. If you change the units of *X* and/or Y, the correlation doesn't change.
- You can switch the values of *X* and *Y* in the data set without changing the correlation.

See the following for an example of calculating a correlation.

EXAMPLE

Q. Find and interpret the correlation for the following data set: (2, 3), (6, 8), (3, 5).

A. Step 1: $\bar{x}$ is 3.67, and $\bar{y}$ is 5.33.

Step 2: The standard deviations are $s_x = 2.08$ and $s_y = 2.52$. (See Chapter 2 if you need help calculating standard deviations.)

Step 3: Taking each X value minus the mean of X times each Y value minus the mean of y gives you $(2-3.67)(3-5.33) = 3.891$; $(6-3.67)(8-5.33) = 6.221$; and $(3-3.67)(5-5.33) = 0.2211$.

Step 4: Sum these results to get 10.33.

Step 5: Divide this total by (s_x times s_y) to get 10.33 divided by $(2.08 * 2.52) = 1.97$.

Step 6: Divide 1.97 by $3-1$ (or 2) to get 0.986. This is the correlation between *X* and *Y* for this example.

Interpretation: This correlation indicates a strong positive linear relationship between *X* and Y. (This corresponds to the scatterplot as well; see the following figure.)

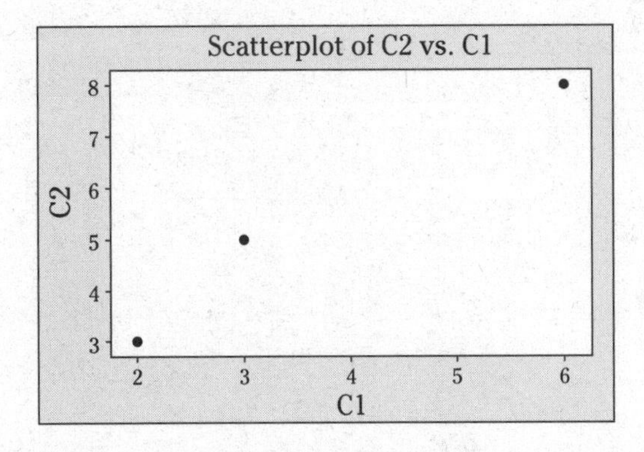

5 Calculate the correlation of the data set (3, 2), (3, 3), (6, 4).

6 Match each of the following correlations to their corresponding scatterplot in the following figure: 0.90, 0.39, −0.74, and 0.57.

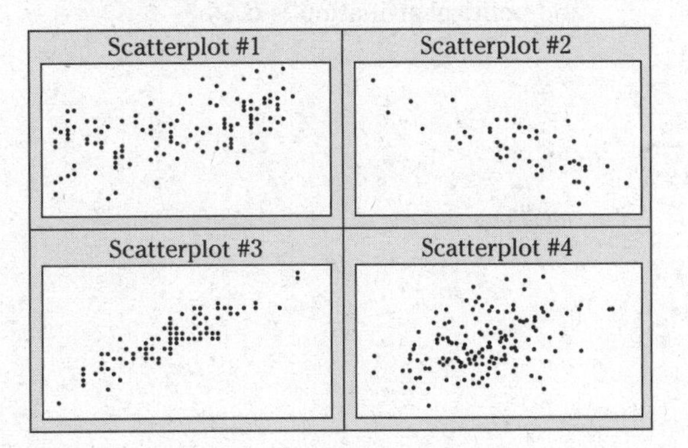

7 Tell whether the following statement is true or false: "The correlation between shoe size and height, in inches, is 0.7. If height is measured in feet, the correlation is $.7 \div 12 = .058$, because 12 inches is equal to 1 foot."

8 Tell whether the following statement is true or false: "The correlation between height and weight is 0.6, so the correlation between weight and height is –0.6."

9 Tell whether the following statement is true or false: "The correlation between gender and political affiliation is 0.65."

10 State whether the following statement is true or false: "The correlation between bushels per acre and annual rainfall is 2.5 inches."

Picking Out the Best Fitting Regression Line

The formula for the *best fitting line* (or *regression line*) is $Y = mX + b$, where m is the slope and b is the Y-intercept. To come up with the best fitting line, you need to find values for m and b so that you have a real equation of a line (for example, $Y = 2X + 3$; or $Y = -10X - 45$).

To save yourself a great deal of time calculating the best fitting line, you should first find the "big five," or five summary statistics that you need for your calculations:

- The mean of the *X* values (denoted $\bar{x}$)
- The mean of the *Y* values (denoted $\bar{y}$)
- The standard deviation of the *X* values (denoted s_x)
- The standard deviation of the *Y* values (denoted s_y)
- The correlation between *X* and *Y* (denoted *r*)

The formula for the *slope* (m) of the best fitting line is $m = r\left(\frac{s_y}{s_x}\right)$, where r is the correlation between X and Y and s_x and s_y are the standard deviations of X and Y, respectively.

To calculate the slope (m) of the best fitting line, you

1. **Take s_y divided by s_x.**
2. **Multiply the result by *r*.**

The formula for the *Y-intercept* (b) of the best fitting line is $b = \bar{y} - m\bar{x}$ where $\bar{x}$ and $\bar{y}$ are the means of X and Y, respectively, and m is the slope.

To calculate the Y-intercept (b) of the best fitting line:

1. **Find the slope (m) of the best fitting line.**
2. **Multiply the slope times $\bar{x}$.**
3. **Take $\bar{y}$ and subtract your result from Step 2.**

TIP

Always calculate the slope before the Y-intercept. The formula for the Y-intercept contains the slope!

See the following for an example of finding the best fitting line by using the "big five" summary statistics.

EXAMPLE

Q. Calculate the equation of the best fitting line for the following small data set: (2, 3), (6, 8), (3, 5).

A. The "big five" summary statistics for this data set are shown in the following table.

Variable	Mean	Standard Deviation	Correlation
X	3.67	2.08	0.986
Y	5.33	2.52	

The slope, *m*, of the best fitting line is $(0.986)(2.52 \div 2.08) = (0.986)(1.21)$, which gives you 1.19. Now to find the Y-intercept (*b*), you take the mean of Y and subtract the mean of X times the slope: $5.33 - (1.19)(3.67) = 0.96$. The best fitting line for this data set is $Y = 1.19X + 0.96$.

11 A coffee barista records the temperature (degrees Fahrenheit) and number of coffees sold for 50 professional football games; you can see the summary of the statistics in the following table. (See the figure from Question 1 for a scatterplot of this data.)

a. Should you use a line to fit this data? Justify your answer.

b. Find the equation of the best fitting regression line.

Variable	Mean	Standard Deviation	Correlation
Temperature (*X*)	35.08	16.29	–0.741
Coffees sold (*Y*)	29,913	12,174	

12 A medical researcher measures bone density and the age of 125 women; you can see the summary of statistics in the following table. (See the figure in Question 2 for a scatterplot of this data.)

a. How well will a line fit this data?

b. Find the equation of the best fitting regression line.

Variable	Mean	Standard Deviation	Correlation
Bone loss (*Y*)	35.008	7.684	0.574
Age (*X*)	67.992	10.673	

13 A golf analyst measures the total score and number of putts hit for 100 rounds of golf an amateur plays; you can see the summary of statistics in the following table. (See the figure in Question 3 for a scatterplot of this data.)

a. Is it reasonable to use a line to fit this data? Explain.

b. Find the equation of the best fitting regression line.

Variable	Mean	Standard Deviation	Correlation
Total score (*Y*)	93.900	7.717	0.896
Putts hit (*X*)	35.780	4.554	

Agricultural scientists try to predict corn production by using annual rainfall. They measure crop yields (bushels per acre) and annual rainfall (inches) for 150 one-acre plots; you can see the summary of statistics in the following table.

a. Identify the *X* variable and the *Y* variable in this problem.

b. Find and try to interpret the slope in the context of corn and rainfall. (More practice on interpretation in the next section.)

Variable	Mean	Standard Deviation	Correlation
Rainfall	47.844	9.38	0.608
Corn	150.77	19.76	

Interpreting the Regression Line and Making Predictions

Any equation or function, such as the best fitting regression line, that you use to estimate or predict a relationship between two variables is called a *statistical model*. Using a model, you can predict the value of *Y* by using *X*. How do you do it? Choose a value for *X*, plug it into the equation, and find the estimated value for *Y*. But understanding what the results mean is just as important as calculating them.

The slope of the regression line is a measure of how much change you expect in *Y* when you alter *X* by one unit. The *Y*-intercept of the regression line is the expected value of *Y* when *X* is equal to 0. The *Y*-intercept may or may not be statistically relevant to the problem because the data collection may not have taken place around the point $X = 0$. However, the *Y*-intercept is mathematically relevant because it marks the place where the line that best fits the data crosses the *Y*-axis.

You can't plug in *any* value for X and accurately predict Y just because you have a model. For example, you can't plug in a number higher than your highest X value in the data set or lower than your lowest X value in the data set. Why not? Because you don't have any data for X in those ranges. Who's to say the line still works outside the data-collection area? You can't plug in extremely low or high values of X and expect the model to work. Making predictions by using X values that fall beyond the range of the data you collect is called *extrapolation*. Extrapolation is a statistical no-no. So when can you use the line to predict Y for a given value of X? When the line fits well, and only when you plug in values for X that lie within the range of where you collect the data.

REMEMBER

One last note about interpreting the regression results: Just because X and Y have a linear relationship and you use values of X to predict values of Y doesn't mean that a change in X *causes* a change in the outcome of Y. It depends on how your data were collected. If the experiment is well designed, you can assume a cause-and-effect relationship. If the experiment is an observational study, you can't claim cause and effect. This presents a big problem statisticians see all the time, and you can count on your instructor asking you about it. (See Chapter 16 for more on observational studies and experiments.)

See the following for an example of deciding which values of X you can make predictions for.

EXAMPLE

Q. Using the figure from the first example problem (from the section "Relating X and Y with a Scatterplot") and the table from the third (from the section "Picking Out the Best Fitting Regression Line"), for what values of X do you feel confident making predictions about Y?

A. 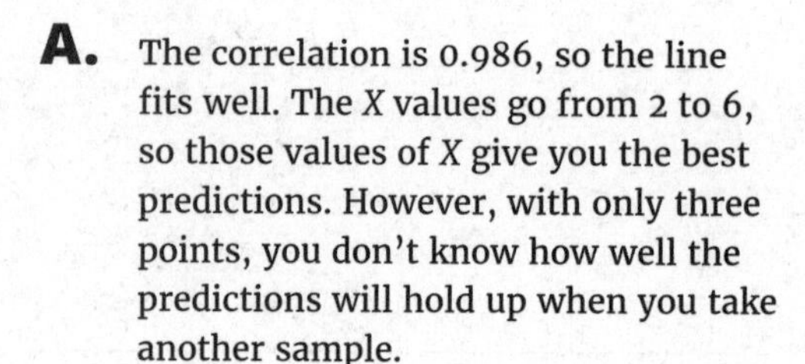
The correlation is 0.986, so the line fits well. The X values go from 2 to 6, so those values of X give you the best predictions. However, with only three points, you don't know how well the predictions will hold up when you take another sample.

15 Answer the following, using the figures and tables from the temperature versus coffee sales data from Questions 1 and 11:

a. How many coffees should the manager prepare to make if the temperature is 32°F?

b. As the temperature drops, how much more coffee will consumers purchase? (*Hint:* Use the slope.)

c. For what temperature values does the regression line make the best predictions?

16 Answer the following, using the figures and tables from the age versus bone loss data in Questions 2 and 12:

a. For what ages is it reasonable to use the regression line to predict bone loss?

b. Interpret the slope in the context of this problem.

c. Using the data from the study, can you say that age causes bone loss?

17 Referring to the figures and tables from the golf data in Questions 3 and 13, what happens as you keep increasing X? Does Y increase forever? Explain.

18 Using the results from the rainfall versus corn production data in Question 14, answer the following:

a. Find and interpret the slope in the context of this problem.

b. Find the Y-intercept in the context of this problem.

c. Can the Y-intercept be interpreted here?

Checking the Fit of the Regression Line

Before you can make any predictions of Y based on X, you need to check to be sure that the regression line you use to make predictions fits the data well. A good fit is a good indicator that after you take the data away and use the model to predict Y for the next X, the model will do a good job.

Here is how you can check the fit of your regression line:

1. **Check the scatterplot to make sure that you see a linear pattern in the data.**
2. **Calculate the correlation and make sure that it's strong enough in either the positive or negative direction.**

   "Strong enough" to most statisticians means generally beyond 0.6 or –0.6, but this is just a general rule.

TIP

You should do Steps 1 and 2 before you fit the regression line. If Steps 1 and 2 don't check out, fitting the regression line is not advised!

3. **Create the regression line and draw it on the scatterplot. Make sure it has the right look and fit.**

   That is, make sure you don't find any places where the line is consistently above or consistently below the data, or situations that indicate the data may have some curvature to it and that a line may not be the best model to fit.

4. **Calculate the value of r^2.**

After you square the value of *r*, you get a value between 0 and 1, which you can interpret as a percentage. You interpret r^2 as the percentage of the variability in the *Y* values that the model between *X* and *Y* can explain.

For example, if you use shoe size to predict foot length, your r^2 value should be pretty high (close to 1), because after you know a person's shoe size, you know almost everything you need to estimate foot length. (However, if you try to use shoe size to predict grade point average, your value of r^2 is very low, meaning you have a lot more explaining to do.)

See the following for an example of discussing how well a regression line fits the data.

EXAMPLE

Q. Using the information from the small data examples (see the example problems in all the previous sections), discuss how well the regression lines fit the data. Also discuss the limitations that having only three data points presents.

A. A scatterplot with only three points doesn't say much. After all, you can fit a line with any two points, so three points doesn't define any real pattern. You could fit a line to these three points and it wouldn't fit badly, but because you have only three data points, you can't be as confident that your line would be the same if you had another sample. You need more data to develop a more credible model.

Note: Always check the sample size when you look at correlations. After all, you can fit a perfect straight line with only two points, but do you get that same data next time?

19 Comment on the fit of the regression line for the bone loss data, using the information from Questions 2 and 12.

20 Comment on the fit of the regression line for the golf data, using the information from Questions 3 and 13.

21 Examine the fit of the line used to predict coffee sales from temperature, using the information from Questions 1 and 11.

a. How much can the concessions manager rely on this model to make predictions about coffee sales? (***Hint:*** Use r^2.)

b. Using only the value of r^2 and the scatterplot, find r.

c. Explain why knowing r^2 isn't enough to find the correlation.

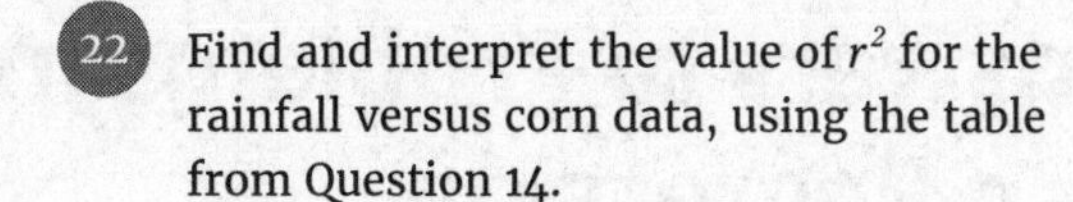

22 Find and interpret the value of r^2 for the rainfall versus corn data, using the table from Question 14.

Answers to Problems in Correlation and Regression

1. You can see a fairly strong negative relationship between temperature and the number of coffees sold. Colder temperatures are associated with more coffees sold, and warmer temperatures are associated with fewer coffees sold.

2. Scatterplots can differ in their appearance in terms of scale, so without the correlation itself, you can't be really specific. However, based on the scatterplot you see here, you notice a weak to moderate positive relationship between age and bone loss (the points are farther from the line in general than in Question 1).

3. You can see a fairly strong positive linear relationship between number of putts and the total score during the golf rounds. (***Note:*** A putt is a stroke that takes place only when you hit on the green and near the hole. Number of putts doesn't count the drives or other shots that take place off the green.)

4. A scatterplot that shows no linear relationship usually shows a big scattering of points plotted every which way, with no apparent pattern or linear relationship at all. Other "oddball" situations include points that form a perfect box around the point (0, 0) on the x, y plane, but such situations are typically figments of an instructor's imagination.

5. $\bar{x}$ is $12 \div 3 = 4$, and $\bar{y}$ is $9 \div 3 = 3$. The standard deviations are $s_x = 1.73$ and $s_y = 1$. Step 3 gives you $(3-4)(2-3)+(3-4)(3-3)+(6-4)(4-3)=3$. Dividing the result by (s_x times s_y) gives you $3 \div (1.73 * 1) = 1.73$. Now divide that result by $3-1$ (or 2) to get 0.865, or 0.87. *Interpretation:* This (small) data set has a strong positive linear relationship between X and Y.

6. The correlations for scatterplots 1–4, in order, are 0.57, –0.74, 0.90, and 0.39.

7. False. Correlation doesn't change if you change the units of X and/or Y.

8. False. Correlation doesn't change if you switch X and Y.

9. False. Correlation doesn't apply to categorical variables; it applies only to quantitative variables. (Even though you assign a number to a categorical variable, those numbers don't mean anything.)

10. False. Correlation is always between –1 and +1, and it is unitless.

11. You can follow the example problem to work through the calculations if you want to.

    a. Yes, the correlation (–0.741) is fairly close to –1, and the scatterplot shows that a line would fit the data well.

TIP

Typically, a correlation at or beyond $\pm.6$ is a pretty good correlation; the closer to ±1 the better, but data sets with a correlation very close to ±1 are few and far between (and possibly suspect).

b. The slope, m, of the best fitting line is $(-0.741)(12{,}174 \div 16.29) = (-0.741)(747.33)$, which gives you –553.77. To find the Y-intercept (b), you take the mean of Y and subtract the slope times the mean of X, which becomes $29{,}913 - (-553.77 * 35.08) = 29{,}913 + 19{,}426.25 = 49{,}339.25$. The best fitting line for the data set is $Y = -553.77X + 49{,}339.25$. The estimated number of coffees sold equals –553.77 times the temperature (in Fahrenheit) +49,339.25.

WARNING

Watch for negative slopes in your calculations; they can create problems if you aren't careful.

(12) This problem resembles what you may read about or see in the media, because it's based on a medical study that relates to health issues; something of great interest to the public.

a. The line doesn't fit the data particularly well; the correlation is only 0.574. However, in most research circles, 0.574 is an acceptable correlation to work with. In the context of this problem, it means that although age is an important factor, other factors influence bone loss as well.

b. The slope, m, of the best fitting line is $(0.574)(7.684 \div 10.673) = (0.574)(0.719)$, which gives you 0.413. To find the Y-intercept (b), you take the mean of Y and subtract the slope times the mean of X, which becomes $35.008 - (0.413 * 67.992) = 35.008 - 28.08 = 6.92$. The best fitting line for the data set is $Y = 0.413X + 6.92$. The estimated amount of bone loss is 0.413 times age plus 6.92.

(13) This problem shows that it's not how hard you hit the ball off the tee; it's the quality of your short game that really matters.

a. The correlation here is 0.896, which shows a moderately strong positive linear relationship. That, added with the fact that the scatterplot shows that the data appear to be linear, means a regression line would provide a fairly reasonable model for this data.

REMEMBER

Don't just look at the correlation to determine whether a line would fit the data well. You also need to examine the scatterplot. If the scatterplot isn't linear, the correlation is meaningless. And you can have a "strong" correlation where the data aren't linear.

When you find correlations only, it doesn't matter which variable you denote X and which you denote Y. However, when you have to find the regression line, it matters a great deal which is which. X is the input variable, the independent variable, and the one that goes into the equation and does the predicting. Y is the output variable, the dependent variable, and the one that X predicts. If you're given a scatterplot in a problem, you have been given a big hint as to what the X and Y variables are. The X variable is on the horizontal (X) axis, and the Y variable is on the vertical (Y) axis.

b. The slope, m, of the best fitting line is $(0.896)(7.717 \div 4.554) = 1.52$. To find the Y-intercept (b), you take the mean of Y and subtract the slope times the mean of X, which becomes $93.9 - (1.52 * 35.78) = 39.51$. The best fitting line for the data set is $Y = 1.52X + 39.51$. The estimated total score is 1.52 times the number of putts, plus 39.51.

(14) If you want to outdo the *Farmer's Almanac*, you need to collect and analyze a ton of data.

a. The key to identifying X and Y (when a scatterplot is not given) is to look at what the researcher is trying to do. (You can't always assume that X is listed first in the statistical information provided.) In the problem, you're told that agricultural scientists try to predict corn production by using annual rainfall. That means they're using annual rainfall to predict corn production. So the X variable is annual rainfall (the variable on which the prediction is based), and Y is the corn production (the variable that provides the outcome you are interested in predicting).

b. The slope, m, of the best fitting line is $(0.608)(19.76 \div 9.38) = 1.28$. The slope of the regression line is *rise over run* — the expected increase in y (corn production) for every one unit increase in X (rainfall). So when the rainfall increases one more inch, the corn production goes up by 1.28 bushels per acre. (***Note:*** You can't tell whether a line would do well here without looking at the scatterplot, as noted in Question 13. The formulas allow you to plug numbers in, however. It's up to you to make the right decision on whether to use those formulas to make good predictions.)

15 What good is doing regression if you can't use it to talk about football?

a. This question translates to finding the expected value of Y when $X = 32$. Plug $X = 32$ into the equation $Y = -553.77X + 49{,}339.25$ (see Answer 11) to get $Y = 31{,}618.61$ coffees. (Better make it 31,619 just to be sure!) This value of Y makes sense if you look at the scatterplot (see the figure in Question 1; after you make a prediction for Y, always look at the scatterplot to see whether the value makes sense).

b. This question basically asks how X and Y are related, which is through the slope. The slope of this line is -553.77, which means that as the temperature goes up, the company sells less coffee. How much less? For every one degree increase in temperature, the company should expect to sell 553.77 fewer cups of coffee on average.

c. The recorded temperatures for the 50 football games ranged from $-10°$ to $70°$F, so you can feel comfortable making predictions for coffee sales if the temperature falls within this range. Outside of it, who knows?

TIP

To interpret any slope, put the value of the slope over 1. As you increase X by one unit, the Y value increases or decreases by whatever the slope is. For example, a slope of 2 means $2 \div 1 = \textit{rise} \div \textit{run}$, so increasing X by 1 is associated with an increase of 2 in Y. A slope of -2 says that increasing X by 1 is associated with a decrease of 2 in Y.

16 Properly interpreting the results of a medical study is very important. (The media often makes mistakes; go figure.)

a. The researcher collects data on women between the ages of 50 and about 85, as you see on the scatterplot, so you can feel comfortable with this range for X.

b. The best fitting line is $Y = 0.413X + 6.92$ (see Answer 12), so the slope is 0.413, or $0.413 \div 1$. For each year women age, their average bone loss increases by 0.413. (Y is the amount of bone loss, not the amount of bone density, which is why the relationship is positive and not negative.)

c. Not necessarily, because the study isn't a controlled experiment. (How could it be?) Other factors may influence bone loss, such as diet (which age also affects). The longer you go without calcium, the more bone loss you may experience, for example. (See Chapter 16 for more information on experiments and observational studies.)

17 You can't assume a player will go on putting forever; everyone has to stop sometime. Therefore, the line doesn't go on forever. Also, you have to putt a minimal amount of times, so the number of putts, for example, on 18 holes is almost always at least 18.

18 You don't have a scatterplot here (oftentimes you don't), but you can still tell plenty from the statistics given.

a. The slope is 1.28, which you calculate in your answer to Question 14. See my Answer 14 for an interpretation.

b. To find the Y-intercept (b), you take the mean of Y and subtract the slope times the mean of X, which becomes $150.77 - (1.28 * 47.84) = 89.53$ bushels.

c. To see whether the Y-intercept is interpretable, you need to look at the scatterplot and see if the scientists collected any data during years without rainfall (in other words, where $X = 0$, do they have any data for Y?). You can safely assume, however, that with zero rainfall, the corn simply wouldn't grow, making the Y-intercept non-interpretable.

TIP

Scatterplots plot the data at their location, and they often don't show everything starting at $X = 0$. You notice this on the corn data scatterplot in the following figure. (Yeah, I decided to break down and show it to you.) Not having the data for $X = 0$ makes it harder to visualize where the line crosses the Y-axis because that part of the graph where $X = 0$ isn't shown. Be aware of this issue when drawing a line on your scatterplot, and don't try to interpret the Y-intercept if it's too far away from where the data were actually collected, because you don't know in that case if a line even still fits in that area.

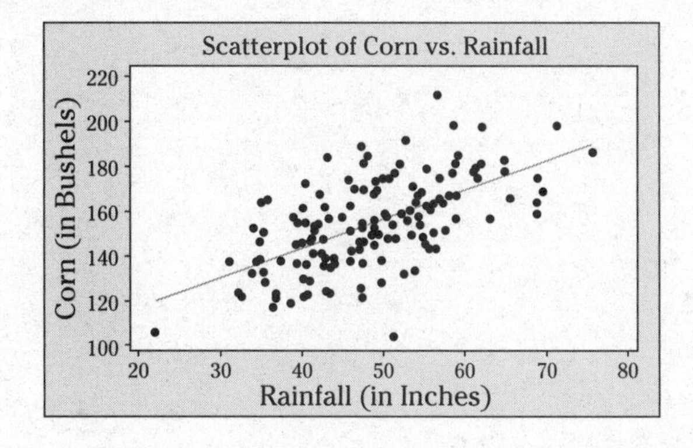

19 The scatterplot shows a "weak to moderate" uphill relationship, and the correlation is indeed weak to moderate, at 0.574. The value of r^2 $(= 0.574^2)$ is only 0.329, or 32.9 percent, indicating that a lot of other variables besides age need to be taken into account to assess bone loss.

20 The golf data has a strong correlation, and the scatterplot shows no problems. The data seem to fall very close to the line. The value of r^2 is $.896^2$, which equals 0.803, or 80.3 percent. So the number of putts explains 80.3 percent of the variability in total score.

21 Situations where the value of r is negative can present special problems of their own.

a. The value of r is -0.741, so squaring it gives you 0.549, or 54.9 percent. So temperature explains 54.9 percent of the variability in coffees sold. The manager can rely on temperature to explain about half the variability in coffee sales, but the rest has to come from other information (time of day, age of the people going to the game, and so on).

REMEMBER

Because r is between -1 and $+1$, squaring it makes the number smaller. An r of 0.7, which is pretty good, gives you only 0.49 after you square it. Just something to be aware of when interpreting r^2.

b. In this case, because you know r^2 is 0.549, you can take the square root and get 0.741. But this isn't the correlation. Why? Because it isn't negative, and you know that it should be negative because you saw it in Question 11. You have to look at the scatterplot (or the sign of the slope of the regression line) to see that the relationship is downhill. Include the minus sign to get $r = -0.741$.

c. If you have the value of r^2 and you need r, you have to check the scatterplot or the slope of the line to see if you need to make r positive or negative after you take the square root.

Part c is a very popular test question. Be on the lookout and study up!

(22) The correlation for rainfall and corn yield is 0.608 (see Question 14). Squaring 0.608 gives you $r^2 = 0.37$, or 37 percent. Therefore, rainfall can explain 37 percent of the variation in corn yield. (Other factors explain the rest of the variability.)

TIP

The value of r^2 is most easily interpreted as a percent because the value is always between 0 percent and 100 percent.

5

The Part of Tens

IN THIS PART . . .

Find ten math tips that point you to a better grade.

Discover ten common statistical mistakes that you should be able to detect and avoid.

Find out my top ten statistical formulas.

Use the tables in the Appendix to solve problems throughout this workbook.

» Being watchful to avoid common math mistakes

» Reviewing your answers with a critical eye

Chapter 19

Math Review: Ten Steps to a Better Grade

Statistics and math are very different subjects, but you use a certain amount of mathematical tools to do statistical calculations. Sometimes you can understand the statistical idea but get bogged down in the formulas and calculations and end up getting the wrong answer. This chapter helps you avoid making the common math mistakes that can cost you points on homework and exams. Read on to increase your confidence with the math tools you need for statistics.

Know Your Math Symbols

The most basic math symbols are +, −, ∗ (multiplication), and ÷ (division); but have you ever seen a ± sign? It means *plus or minus* and indicates a lower bound and an upper bound for your answer. Other commonly used math symbols involve the Greek letter "capital" sigma, which stands for *summation*. Used here, $\sum_{i=1}^{n} x_i$, it means add all the indexed numbers from the lowest index (under the summation sign) to the highest index (shown above the summation sign).

In math formulas, you often leave out the ∗ sign for multiplication; for example, $2\frac{s}{\sqrt{n}}$. Suppose that you need to solve that equation when $s = 2$ and $n = 16$. This expression means two times the quantity (s divided by the square root of n). When s equals 2 and n equals 16, you have $2\frac{s}{\sqrt{n}} = 2 * \frac{2}{\sqrt{16}} = 2 * \frac{2}{4} = 1$.

REMEMBER

If you come across a math symbol that you don't understand, ask for help. You can never get comfortable with that symbol until you know exactly what you use it for and why. You may be surprised that after you lift the mystique, math symbols aren't really as hard as they seem to be. They simply provide you with a shorthand way of expressing something that you need to do.

Uproot Roots and Powers

Remember that squaring a number means multiplying it by itself two times, not multiplying by two. And taking the square root means finding the number whose square gives you your result; it doesn't mean dividing the number by 2. Using math notation, x^2 means square the value (so for $x = 3$, you have $3^2 = 9$); and $\sqrt{x}$ means take the square root (for $x = 9$, this means the square root of 9 is 3).

You can't take the square root of a negative number, because you can't square anything to get a negative number back. So anything under a square root sign has to be a nonnegative quantity (that is, it has to be greater than or equal to 0).

These ideas may seem straightforward, but like everything else, they can get complex very fast. If you need to find the square root of an entire expression — for example, $\sqrt{2 + 2(7.1)}$ — put everything under the square root sign in parentheses so your calculator knows to take the square root of the entire expression, not just part of it. In this case, you get the square root of $[2 + 2(7.1)]$, which is the square root of 16.2, which gives you approximately 4.02.

Statistics often deal with percentages — numbers that in decimal form are between 0 and 1. You need to know that numbers between 0 and 1 often act differently than large numbers do. For example, numbers larger than 1 get smaller when you take the square root, but numbers between 0 and 1 get larger when you take the square root. For example, the square root of 4 is 2 (which is smaller than 4), but the square root of $\frac{1}{4}$ is $\frac{1}{2}$ (which is bigger than $\frac{1}{4}$). And when you take powers, the opposite happens. Numbers larger than 1 that you square get larger; for example, 3 squared is 9 (which is larger than 3). Numbers between 0 and 1 that you square get smaller; for example, $\frac{1}{3}$ squared is $\frac{1}{9}$ (which is smaller than $\frac{1}{3}$).

Treat Fractions with Extra Care

Every fraction contains a top (numerator) and a bottom (denominator). For example, in the fraction $\frac{3}{7}$, 3 is the numerator and 7 is the denominator. But what does a fraction really mean? It means division. The fraction $\frac{3}{7}$ means take the number 3 and divide it by 7.

WARNING

A common mistake is to read fractions upside down in terms of what you divide by what. The fraction $\frac{1}{10}$ means 1 divided by 10, not 10 divided by 1. If you can hold on to an example like this that you *know* is correct, it can stop you from making this mistake again later when the formulas get more complicated.

Complex fractions contain another fraction either in the numerator, the denominator, or both. They are also known as a fraction within a fraction. The key to dealing with complex fractions is proper use of parentheses. For example, suppose that you need to find and describe what the complex fraction $\frac{24-20}{\frac{4}{\sqrt{100}}}$ means. The denominator is its own fraction, and you can figure it separately. Or if you figure it all at once, put the numerator in parentheses so your calculator knows that it's that entire part, and divide it all by the denominator (which also should be in parentheses if it's its own fraction). In this case, the answer is $(24-20)=4$ on top and 4 divided by the square root of 100 (10), which is 0.4 on the bottom. Top divided by bottom gives you $4 \div .4 = 10$.

What happens if you don't put parentheses around the denominator when you need to? Consider the simple fraction $\frac{2}{4+6}$. It may seem like you should take 2 divided by 4 and then add 6 to get $2 \div 4 + 6 = 0.5 + 6 = 6.5$. But the correct answer is 2 divided by the quantity in the denominator — in other words, 2 divided by 10, which is 0.2. Your calculator needs to know that the denominator is all together, so you need to put parentheses around it and type in $2 \div (4+6)$ to get it right. The fraction $\frac{2+4}{6}$ looks similar, but it, too, gives a different answer, again because of the way you have to group the numbers during your calculations. This equation says to take $2+4$ first, which equals 6, and then divide that by 6 to get $6 \div 6 = 1$. To calculate this correctly with a calculator, you need to put parentheses around the entire numerator, so type in $(2+4) \div 6$ to get it right.

REMEMBER

To safely move expressions from your book to your calculator, you need to put parentheses in to indicate what each part of the fraction is.

Obey the Order of Operations

To follow the order of math operations, remember "PEMDAS": Parentheses, Exponents (powers of a number), Multiplication and Division (interchangeable), and Addition and Subtraction. Failing to follow the order of operations can result in a big mistake.

TIP

To remember the letters in PEMDAS for the order of operations, try this: "Please Excuse My Dear Aunt Sally."

Suppose, for example, that you need to calculate the following: $(-6+5+\frac{1}{2}-8+10) \div 5$. First, calculate what's in parentheses. You can either type it just as it looks into your calculator or do $\frac{1}{2} = 0.5$ separately and then plug it in as $-6+5+0.5-8+10$. You should get $\frac{3}{2}$ or 1.5. Next, divide by 5 to get $(3 \div 2) \div 5$ or $1.5 \div 5$, which equals 0.3.

As another example, suppose that you plug numbers into a formula, and the result is $5.56 \pm 1.96\sqrt{\frac{10.6}{200}}$. You need to simplify this to get your final answer. Remembering the order of operations (PEMDAS), you find out what's under the square root first (remember, the square root is an exponent, because you can write it as the power $\frac{1}{2}$). So you take $10.6 \div 200 = .053$ and find the square root, which is 0.230. You take 0.230 and multiply by 1.96 to get 0.451. Now you need to find $5.56 \pm .451$. First, take $5.56 - 0.451$, which equals 5.109. Next, take $5.56 + 0.451 = 6.011$. Your final answer is (5.109, 6.011).

Suppose that you have to calculate the following: $\frac{(4-2)}{5}$, and $\frac{(-2-3)}{5}$. For the first term, do the parentheses on top first to get $4 - 2 = 2$. The denominator is 5, so you take the numerator (2) divided by the denominator (5) to get $2 \div 5 = 0.40$. The sign is very important here (as it is with all formulas). Now the second term. Suppose that you want to calculate the second term all at once instead of taking two steps. Type it into your calculator exactly like this: $(-2-3) \div 5$. You get −1.

WARNING

In the previous example, the parentheses are critical around the numerator. Without them, you get $-2-3 \div 5$, which is $-2-.6$, which equals −2.6 (the wrong answer).

How would you calculate this expression: $\frac{(2-4.3)(4-6.1)}{(6.32)(1.12)}$? One way to do this calculation is to figure out the calculations for the numerator and denominator separately and then divide them. Take the numerator first and figure items in parentheses together as a unit. In other words, take $(2-4.3)*(4-6.1)$ to get $(-2.3)*(-2.1)$, which is 4.83. The denominator is $6.32*1.12 = 7.078$. Take the numerator divided by the denominator, which is $4.83 \div 7.078 = .68$. Or you can put the entire expression into your calculator at once. Just remember to put a separate set of parentheses around the entire numerator and one around the entire denominator so your calculator knows which part is which. You type it in like this: $[(2-4.3)*(4-6.1)] \div (6.32*1.12)$.

Avoid Rounding Errors

Rounding errors can seem small, but they can really add up — literally. Many statistical formulas contain several different types of operations that you can do either all at once, using parentheses properly, or separately, as many students elect to do. Doing the operations separately and writing them down with each step is fine, as long as you don't round off numbers too much at each stage.

For example, suppose that you have to calculate $1.96\frac{5.2}{\sqrt{200}}$, and you want to write down each step separately rather than calculate the equation all at once. Suppose that you round off to one digit after the decimal point on each calculation. (I don't advocate doing this; I just want to show the impact of that rounding.) First, you take the square root of 200 (which rounds to 14.1), and then you take 5.2 divided by 14.1, which is 0.369; you round this to 0.4. Next, you take 1.96 times 0.4 to get 0.784, which you round to 0.8. The actual answer, if you do all the calculations at once with no rounding, is 0.72068, which safely rounds to 0.72. What a huge difference! What would this difference cost you on an exam? At worst, your professor would reject your answer outright, because it strays too far from the correct one. At best, he would take off some points, because your answer isn't precise enough.

Instead of rounding to one digit after the decimal point, suppose that you round to two digits after the decimal point each time. This still gives you the incorrect answer of 0.73. You've come closer to the correct answer, but you're still technically off, and points may be lost. Statistics is a quantitative field, and teachers expect precise answers. What should you do if you want to do calculation steps separately? Keep at least two significant digits after the decimal point during each step, and at the very end, round off to two digits after the decimal point.

REMEMBER

Don't round off too much too soon, especially in formulas where many calculations are involved. Your best bet is to use parentheses and use all the decimal places in your calculator. Otherwise, keep at least two significant digits after the decimal point until the very end.

Get Comfortable with Formulas

Don't let formulas get in your way. Think of them as mathematical shorthand. Suppose that you want to find the average of some numbers. You sum the numbers and divide by *n* (the size of your data set). If you have only a few numbers, writing out all the instructions is easy, but what if you have 1,000 numbers? Mathematicians have come up with formulas as a way of saying quickly what they want you to do, and the formulas work no matter the size of your data set. The key is getting familiar with formulas and practicing them.

For example, suppose that you have to find $\frac{\sum_{i=1}^{n} x_i}{n}$, where $x_1 = 2$, $x_2 = 4$, $x_3 = 6$, and $x_4 = 8$. The x_i in the formula means "the *i*th *x* value," and *i* starts at 1 and goes up to *n* (the size of the data set). Here, $n = 4$ is the place where it stops. This formula is shorthand for finding the average of the four values. It says to take the first value, x_1, add the next value to it (x_2), add the next value to it (x_3), and keep going until you add all the numbers together (in this case, x_4 is the last one). After all that summing, divide by *n*, which here is four. (Notice that 4 is also the number at the top of the summation sign, for the same reason.) This gives you $(2+4+6+8) \div 4 = 20 \div 4 = 5$.

WARNING

The parentheses around the numerator values of a fraction are important. If you don't use parentheses correctly in the last example, you get $2+4+6+8 \div 4 = 2+4+6+2 = 14$ (the wrong answer).

Now, suppose that you have to take the numbers $x_1 = -1$, $x_2 = 0$, $x_3 = 1$, and calculate each of the following: $\sum_{i=1}^{n} x_i^2$ and $\left(\sum_{i=1}^{n} x_i\right)^2$. Your first question should be: What's the difference between these two formulas? The first formula is the sum of the squares of the *x* values. In other words, square each *x* value and add it to the next one, and keep going until you hit all the numbers. This gives you $(-1)^2 + 0^2 + 1^2$, which equals $1+0+1=2$. The second formula is the square of the sum of the *x* values. You add them all up first and then square the whole thing. So you take $-1+0+1=0$ and then square it to get 0.

Stay Calm When Formulas Get Tough

Suppose that you encounter a formula that's a little more involved. How do you remain calm and cool? By starting with small formulas, learning the ropes, and then applying the same rules to the bigger formulas. That's why you need to understand how the "easy" formulas work and be able use them as formulas; you shouldn't just figure them out in your head, because you don't need the formula in that case. The easy formulas build your skills for when things get tougher.

For example: $\frac{\sum_{i=1}^{n}(x_i - \bar{x})^2}{n-1}$, for $x_1 = 1$, $x_2 = 2$, and $x_3 = 3$. Think about what the formula asks you to do with the numbers before you plunge in and try to do the calculations. This formula wants you to take the first x value, subtract the mean, square it, and then go to the next x value and do the same thing; add all the results together for the three values of x and divide by $3-1$ (because you have $n = 3$ numbers) to get your final answer. In this case, the mean $(1+2+3) \div 3 = 2$. This gives you $\frac{(1-2)^2 + (2-2)^2 + (3-2)^2}{(3-1)}$, which is $(1+0+1) \div 2 = 2 \div 2 = 1$ — the formula for sample variance.

The formula for sample standard deviation is $\sqrt{\frac{\sum_{i=1}^{n}(x_i - \bar{x})^2}{n-1}}$. How do you calculate this?

First, you have to calculate what's under the square root sign, using the previously mentioned steps. Notice that the part under the square root sign is the same as the formula for variance. Now, take the square root. It can be easy to forget the square root at the end when you have so many steps to do, so always double-check the last step. (*Note:* If you've just calculated the sample variance in a previous problem, simply take the square root of the variance to find the standard deviation.)

How should you calculate the following formula: $\frac{(x - \bar{x})}{\frac{s}{\sqrt{n}}}$, for $x = 8$, $\bar{x} = 7.6$, $s = 15.4$, and $n = 100$?

Plug in the values for all the variables, which gives you the expression $\frac{(8-7.6)}{\frac{15.4}{\sqrt{100}}}$. Do the parentheses in the numerator to get $8 - 7.6 = 0.4$. The denominator is 15.4 divided by the square root of 100, which is $15.4 \div 10 = 1.54$. So the final answer is $0.4 \div 1.54 = .26$.

Now you're ready to really take it up a notch. Suppose that you have three pairs of (x, y) values — (1, 10), (2, 20), and (3, 30) — and you have to take those values and find $\sum_{i=1}^{n}(x_i - \bar{x})(y_i - \bar{y})$. (Note that $n = 3$, because you have three $[x, y]$ points.)

1. **The (x_i, y_i) pairs are denoted (x_1, y_1), (x_2, y_2), and (x_3, y_3). Note that the subscript *i* tells you which pair you're looking at.**

   This formula involves $\bar{x}$ and $\bar{y}$ — the mean of the x values and the mean of the y values, respectively.

2. **First, find the means of the x values and y values. For this example, you get $(1+2+3) \div 3 = 2$ and $(10+20+30) \div 3 = 20$.**

3. **Take the first x value and subtract its mean: $(1-2) = -1$.**

4. **Now take its corresponding y value and subtract its mean: $(10-20) = -10$.**

5. **Next, multiply those two values together to get $(-1) * (-10) = 10$.**

6. **Do the same thing for the 2nd pair of (x, y) values. Take x minus its mean $(2-2)$ and y minus its mean $(20-20)$, and multiply them to get $0 * 0 = 0$.**

7. **Now, do it again for the third set of (x, y) values to get $(3-2)*(30-20)=1*10=10$.**
8. **Add all the results together to get $10+0+10=20$.**

You've just worked through a part of the formula for correlation, which you can examine more in-depth in Chapter 18.

Feel Fine about Functions

Many times in math and statistics, different variables are related to each other. For example, to get the area of a square, you take the length of one of the sides and multiply it by itself. In mathematical notation, the formula looks like this: $A=s^2$. This formula really represents a function. It says that the area of the square depends on the length of its sides. It also means that all you have to know is the length of one of the sides to get the area of the square. In math jargon, you say that the area of a square is a function of the length of its sides. *Function* just means "depends on."

Suppose that you have a line with the equation $y=2x+3$. The equation conveys that x and y are related, and you know how they're related. If you take any value of x, multiply it by two and add three, you get the corresponding value for y. Suppose that you want to find y when x is –2. To find y for a given x, plug in that number for x and simplify it. In this case, you have $y=(2)(-2)+3$. This simplifies to $y=-4+3=-1$.

You can also take this same function and plug in any value for y to get its corresponding value for x. For example, suppose that you have $y=2x+3$, and you're given $y=4$ and asked to solve for x. Plugging in 4 for y, you get $4=2x+3$. The only difference is, you normally see the unknown on one side of the equation and the number part on the other. In this case, you see it the other way around. Don't worry about how it looks; remember what you need to do. You need to get x alone on one side, so use your algebra skills to make that happen. In this case, subtract 3 from each side to get $4-3=2x$, or $1=2x$. Now divide each side by 2 to get $0.5=x$. You have your answer.

REMEMBER

You can use a formula in many different ways. If you have all the other pieces of information, you can always solve for the remaining part, no matter where it sits in the equation. Just keep your cool and use your algebra skills to get it done.

Certain commonly used functions have names. For example, an equation that has one x and one y is called *a linear function*, because when you graph it, you get a straight line. Statistics uses lines often, and you need to know the two major parts of a line: the slope and the y-intercept. If the equation of the line is in the form $y=mx+b$, m is the slope (the change in y over change in x), and b is the y-intercept (the place where the line crosses the y-axis). Suppose that you have a line with the equation $y=-2x-10$. In this case, the y-intercept is –10, and the slope is –2.

TIP

The slope is the number in front of the x in the equation $y=mx+b$. If you rewrite the previous equation as $y=-10-2x$, the slope is still –2, because –2 is the number that goes with the x. And –10 is still the y-intercept.

Know When Your Answer Is Wrong

You should always look at your answer to see whether it makes sense, in terms of what kind of number you expect to get. Can the number you're calculating be negative? Can it be a large number or a fraction? Does this number make sense? All these questions can help you catch mistakes on exams and homework before your instructor does.

The formula for the variance of a data set is $\frac{\sum_{i=1}^{n}(x_i - \bar{x})^2}{n-1}$. Suppose that you do the calculations on your data set and your variance turns out to be a negative number. Your answer can't possibly be right. Why? The numerator of the variance formula says to square each value, which results in a number greater than or equal to 0. You divide by n, which is the size of the data set, so the number can never be negative (< 0). Your result has to be greater than or equal to 0.

Suppose that your data set is $x_1 = 100$, $x_2 = 150$, $x_3 = 125$, and $x_4 = 110$. You calculate the average (mean), using the formula $\frac{\sum_{i=1}^{n} x_i}{n}$, and get a negative number. Your old pal Bob gets the number 151.6 for his answer. Your friend Sue gets $100 + 150 + 125 + 110 \div 4 = 402.5$. Each of these answers has to be incorrect. Why? Your answer is wrong, not because an average can't be negative, because it can be negative sometimes (look at the average of -1, -2, and -3), but because all your data is positive, the average, which simply sums up the data and divides by n (a positive value), has to be positive. Bob, on the other hand, gets a value outside the range of the data. When you take the average of a group of numbers, the result has to be somewhere between the smallest and largest numbers in the data set. Because of what Bob found out, you know Sue's answer is also wrong, but why? Because Sue didn't put parentheses around the entire sum $(100 + 150 + 125 + 110)$ before dividing by 4. Those parentheses are really important!

Suppose that you're working with fractions, and you want to find $^{816}/_{200}$. You can't remember if you should take 816 divided by 200 or take 200 divided by 816. Which way do you go and why? Should your final answer be less than or greater than 1? When working with fractions, you always take the numerator and divide it by the denominator; so you should divide 816 by 200 to get 4.08. This number is larger than 1. First, you know it has to be greater than 0 because it's positive. Then, because the numerator (top) is larger than the denominator (bottom), that means the entire number must be greater than 1. ***Note:*** If this was supposed to be a probability, it would be wrong because all probabilities must be between 0 and 1.

TIP

In any fraction, if the numerator (top) is larger than the denominator (bottom), the result is greater than 1. If the numerator (top) is smaller than the denominator (bottom), the result is less than 1. And if the numerator (top) and denominator (bottom) are exactly equal, the result is exactly 1.

One of the formulas people use a great deal in statistics is $\bar{x} \pm 1.96\frac{\sigma}{\sqrt{n}}$. The sample size is n. If you increase n, what happens to the number after the $\pm$ sign? Does it get larger, smaller, or does it not even depend on n? The letter n is the sample size (so $n \geq 1$), and it sits in the denominator of a fraction under a square root. Because it sits in the denominator of a fraction, increasing n increases the denominator, which makes the entire fraction smaller. So the part after the $\pm$ sign decreases. If you make n smaller, the opposite happens because n is in the denominator.

Show Your Work

You see the instructions "Show your work!" on your exams, and your instructor harps and harps on it, but still, you don't quite believe that showing your work can be that important. Take it from a seasoned professor, it is. Here's why:

- **Showing your work helps the person grading your paper see exactly what you tried to do, even if the answer is wrong.** This works to your advantage if your work was on the right track. The only way to get partial credit for your work is to show that you had the right idea, and you must do this in writing.
- **Not showing your work makes it hard on the person grading your paper and can cost you points in an indirect way.** Grading is a tremendous amount of work. Now before I ask you to feel sorry for your teachers, which I doubt I can talk you into, let me show you how the "grading effect" on your teacher ultimately affects you. Your teacher has a big pile of papers to grade and only so much time (and energy) to grade them all. A paper with a big messed up area of scribbling, erasing, crossing out, and smudging rears its ugly head. It has no clear tracks as to what's happening or what the student was thinking. Numbers are pushed around every which way with no clear cut steps or pattern to follow. How much time can (will) teachers spend trying to figure out this problem? Teachers have to move on at some point; we can only do so much to try to figure out what students were thinking during an exam.

  Here's another typical situation. A teacher looks at two papers, both with the right answer. One person wrote out all the steps, labeled everything, and circled the answer, but the other person simply wrote down the answer. Do you give both people full credit? Some teachers do, but many don't. Why? Because the instructor isn't sure whether you did the work yourself. Teachers don't typically advocate doing math "in your head." We want you to show your work, because someday, even for you, the formulas will get so complicated that you can't rely on your mind alone to solve them. Plus, you do need to show evidence that the work is your own.

  What if you write down the answer, and the answer is wrong, but only a tiny little mistake led to the error? With no tracks to show what you were thinking, the teacher can't give you partial credit, and the littlest of mistakes can cost you big time.
- **Showing your work establishes good habits that last a lifetime.** Each time you work a problem, whether you're working in class, on homework, to study for an exam, or on an exam, if you follow the same procedure each time, good things will happen.

TIP

Here's what I always do when I work a math-related statistics problem; if it works for me and my students, I'm betting it can also work for you:

1. **Write out the formula you plan to use, in its entirety (letters included).**
2. **Clearly write down what number you plug in for each variable in the formulas; for example, $x = 2$ and $y = 6$.**
3. **Work out the calculations in a step-by-step manner, showing each step clearly.**
4. **Circle your final answer clearly.**

The biggest argument students give me for not showing their work is that it takes too much time. Yes, showing your work takes a little more time in the short run. But I argue that showing your work actually saves time in the long run, because it helps you organize your ideas clearly the first time, cuts down on the errors you make the first time around, and lessens your need to have to go back and double check everything at the end. If you do have time to double-check your answers, you have an easier time seeing what you did and finding a potential mistake. Trust me, showing your work is a win-win situation. Try showing your work a little more clearly, and see how it impacts your grades.

formulas

» Practicing the calculations

» Understanding the properties and behavior of these formulas

» Checking your answers and avoiding common mistakes

Chapter 20
Top Ten Statistical Formulas

This chapter reviews my top ten statistical formulas and the steps and tips for calculating them. I provide a nice last-minute review/refresher to see what you remember and to show you what you need to spend a little more time on. You also get some tips to help you check your answers and to let you know what to expect your results to look like and why.

Mean (or Average)

The *mean*, or the average of a data set, is one way to measure the center of a numerical data set. The notation for the mean is $\bar{x}$. The formula for the mean is $\bar{x} = \frac{\sum x}{n}$, where x represents each of the values in the data set.

To calculate the mean, you

1. **Add up all the numbers in the data set.**
2. **Divide by *n* (the amount of numbers in the data set).**

For example, to find the mean of the numbers 6, 10, and 20, take $6 + 10 + 20 = 36$, and $36 \div 3 = 12$.

What about the mean of the data set 1, 1, 1, 2, 3, 4? Do you count the 1 only once and take $(1+2+3+4) \div 4$? No. You need to count each number in the mean, whether the numbers are distinct or not. So the mean of the data set 1, 1, 1, 2, 3, 4 is $12 \div 6 = 2$.

The mean can be negative if some of the data are negative, and the mean can be 0 if all the negatives average out with all the positives.

Suppose that you calculate the mean of the data set 1, 2, 5, 5, 4, 4, 2, 6 and get 8.2. Is your answer right? No. The mean has to be somewhere between the largest and smallest numbers in the data set.

What effect does an outlier have on the mean? The mean of the data set 1, 2, 3, 4, 5 is 3. If you add an outlier like 1,000, the mean changes to $(1+2+3+4+5+1{,}000) \div 6 = 169.17$. The outlier really drives the mean upward.

For more practice with the mean, see Chapter 2.

REMEMBER

The mean of a data set has to be somewhere between the largest and the smallest values in the data set. It can be negative, positive, or 0. And the mean is sensitive to outliers.

Median

The *median* (refer to Chapter 2) of a numerical data set is another way to measure the center. The median is the middle value after you order the data from smallest to largest. It doesn't have a commonly recognized formula.

To calculate the median, go through the following steps:

1. **Order the numbers from smallest to largest.**
2. **For an odd amount of numbers, choose the one that falls exactly in the middle. You've pinpointed the median.**
3. **For an even amount of numbers, take the two numbers exactly in the middle and average them to find the median.**

For example, the median of the data set 2, 10, 3 is 3. The ordered data set is 2, 3, 10. The middle value is 3. Notice that the median is one of the values in the data set. The data set 10, 2, 3, 5 has an even number of values. The ordered set looks like 2, 3, 5, 10. The median is $(3+5) \div 2 = 4$. Notice that the median isn't one of the values in the data set.

What about the median of the data set 1, 1, 1, 2, 3, 4? Do you count the 1 only once and take the number in the middle of 1, 2, 3, 4? No. You need to count each number in the median, whether the numbers are distinct or not. So the median of the data set 1, 1, 1, 2, 3, 4 is $(1+2) \div 2 = 1.5$.

The median can be negative if some of the data are negative, and the median can be 0 if the data set has an even amount of numbers and the two middle numbers are the same number with opposite signs — like −10 and 10, because $(-10+10) \div 2 = 0$.

Suppose that you calculate the median of a data set. How can you tell whether your answer is right? The median has to be somewhere between the largest and smallest numbers in the data set. If all the numbers in the data set are equal, the median is equal to that value.

What effect does an outlier have on the median? The median of the data set 1, 2, 3, 4, 5 is 3. If you add an outlier like 1,000 to the set, the median is now the middle number of 1, 2, 3, 4, 5, 1,000, which is $(3+4)\div 2 = 3.5$. The addition of 1,000 doesn't affect the median much at all.

REMEMBER

The median is a measure of center. Its value is between the largest and smallest numbers in the data set. It may or may not be one of the data values itself. And it isn't sensitive to outliers (like the mean is), which is a plus. In statistical lingo, the median is said to be resistant to outliers.

Sample Standard Deviation

The *standard deviation* (refer to Chapter 2) of a sample is a measure of the amount of variability in the sample. You can think of it, in general terms, as the average distance from the mean. The formula for the standard deviation is $s = \sqrt{\frac{\sum (x-\bar{x})^2}{n-1}}$.

To calculate the standard deviation, you

1. **Find the average of all the numbers, $\bar{x}$.**
2. **Take each number and subtract the average from it.**
3. **Square each of the resulting values.**
4. **Add them all up.**
5. **Divide by $n-1$.**
6. **Take the square root.**

For example, find the standard deviation of a sample of three values: 0, 10, 11.

1. **Find the mean: $21 \div 3 = 7$.**
2. **Subtract the mean from each number to get $(0-7) = -7$, $(10-7) = 3$, and $(11-7) = 4$.**
3. **Square each of these numbers to get 49, 9, and 16.**
4. **Add these up to get 74.**
5. **Divide 74 by 2 ($3-1$) to get 37.**
6. **Finally, take the square root of 37, which is 6.08.**

Notice that you include the 0 in the calculations; 0 is a number just like all the others, and it deviates from the mean.

REMEMBER

The sample standard deviation is denoted by s, and the population standard deviation is denoted by σ. To get the population standard deviation, use the same calculation as you do for standard deviation, except you divide by N, the total population size.

What kind of number can the standard deviation be? Because you square each of the differences, all the numbers are greater than or equal to 0. Summing those numbers and dividing by $n-1$ (which is positive) gives you a number greater than or equal to 0. So, the standard deviation can never be negative.

What situation gives you the smallest possible standard deviation (0) when all the numbers are the same, like 3, 3, 3, 3? In this case, the mean is 3, the differences are all $3-3=0$, the squares are all 0, summing them gives you 0, and dividing by $4-1$ still gives you 0. It makes sense, because you measure the average distance from the mean. If all the numbers are the same, the average distance from the mean is 0.

What effect does an outlier have on the standard deviation? The standard deviation of the data set 1, 2, 3 is 1. If you add 1,000 to this data set and recalculate the standard deviation, you get 499! Obviously a big difference. You see a large difference because the mean is larger, the differences from the mean are larger, and squaring them makes them even larger.

REMEMBER

The standard deviation can't be a negative number. The smallest it can be is 0, and that happens only if all the data are exactly the same. The standard deviation is sensitive to outliers. Don't forget to take the square root at the very end!

Correlation

Sample *correlation* (refer to Chapter 18) is a measure of the strength and direction of the linear relationship between two quantitative variables *x* and *y*. It doesn't measure any other type of relationship, and it doesn't apply to categorical variables. The formula for correlation is $r=\frac{1}{n-1}\sum\frac{(x-\bar{x})(y-\bar{y})}{s_x s_y}$.

To calculate the correlation (otherwise known as *r*, the sample correlation), you

1. **Find the mean of all the x values and call it $\bar{x}$. Find the mean of all the y values and call it $\bar{y}$.**
2. **Find the standard deviation of all the x values and call it s_x. Find the standard deviation of all the y values and call it s_y.**
3. **For each (x, y) pair in the data set, take x minus $\bar{x}$ and y minus $\bar{y}$ and multiply them together.**
4. **Add all these products together to get a sum.**
5. **Divide the sum by $s_x * s_y$.**
6. **Divide the result by $n-1$, where n is the number of (x, y) pairs. (This is the same as multiplying by 1 over $n-1$.)**

For example, suppose that you have the data set (3, 2), (3, 3), and (6, 4). Follow the steps to find the correlation:

1. **Find the means: $\bar{x}$ is $12 \div 3 = 4$, and $\bar{y}$ is $9 \div 3 = 3$.**
2. **Find the standard deviations: $s_x = 1.73$ and $s_y = 1$.**
3. **Now, multiply: $(3-4)(2-3) = (-1)(-1) = 1$; $(3-4)(3-3) = (-1)(0) = 0$; and $(6-4)(4-3) = (2)(1) = 2$.**
4. **Sum these values to get $1 + 0 + 2 = 3$.**
5. **Divide 3 by (s_x times s_y) or 3 divided by $(1.73 * 1) = 1.73$.**
6. **Divide 1.73 by $3 - 1$ (or 2) to get 0.865, or 0.87.**

REMEMBER

Correlation measures the strength and direction of the linear relationship between two quantitative variables (only). To calculate it, first you calculate the means and standard deviations of *x* and *y*. The correlation *r* is always a number between –1 and +1. (Close to 0 means no linear relationship; close to +1 means a strong positive linear relationship; close to –1 means a strong negative linear relationship.) The correlation is unitless (which means it has no units), and switching the roles of *x* and *y* doesn't change it. Neither does changing the units of *x* and *y*. The correlation is sensitive to outliers.

Margin of Error for the Sample Mean

The *margin of error for your sample mean*, $\bar{X}$ (refer to Chapter 10), is the amount you expect the sample mean to vary from sample to sample. The formula for the margin of error for $\bar{X}$, dealing with samples of size 30 or more, is $\pm Z^* * \frac{\sigma}{\sqrt{n}}$, where Z^* is the standard normal value for the confidence level you want.

To calculate the margin of error for $\bar{X}$, follow these steps:

1. **Determine the confidence level and find the appropriate Z^*.**
2. **Find the standard deviation, σ, and the sample size, n.**
3. **Multiply Z^* by σ divided by the square root of n.**

For example, suppose that you want to calculate the average amount of money spent on a football ticket for a certain top-10 football team. You take a random sample of 100 people, and you find the average price is $55 per ticket with a population standard deviation of $15. You want to estimate the average for all tickets sold for this team and be 95 percent confident in your results. What's the margin of error? Take 15 divided by the square root of 100 (which is 10) to get 1.5. Now multiply 1.5 by whatever Z^* value you need to be 95 percent confident (1.96; see Chapter 11). So the margin of error is $1.96(1.5) = 2.94$. Notice that you never use the $55, because margin of error doesn't have anything to do with the actual sample mean, just its level of precision.

Suppose that a sample of 100 people who root for another team has a mean of $55, but the standard deviation is $30. How does this margin of error compare with the previous one? This margin is larger, because the standard deviation (in the numerator of margin of error) is larger. Calculating it, you get 1.96 times 30 divided by the square root of 100, which is $1.96(3) = 5.88$. If you want to reduce the margin of error here, take a larger sample, because a larger sample has the opposite effect (because n is in the denominator of the margin of error).

For samples of size 30 or less, use the corresponding value from the t-distribution rather than from the standard normal (Z) distribution (see Chapter 8).

REMEMBER

The margin of error is plus or minus (±) the amount calculated in Steps 1 through 3. It follows the ± sign. More confidence and/or more variability in the population means the margin of error increases; larger samples decrease the margin of error.

Sample Size Needed for Estimating μ

If you want to calculate a confidence interval for the population mean with a certain margin of error, you can figure out the sample size you need before you collect any data. The formula for the sample size for estimating μ is $n = \left(\frac{Z^* * s}{MOE}\right)^2$ where Z^* is the standard normal value for the confidence level, MOE is your desired margin of error, and s is the standard deviation. Because s is an unknown that you need, you have to do a pilot study (small experimental study) to come up with a guess for the value of the standard deviation. If you do know the population standard deviation, σ, use it.

To calculate the sample size for estimating μ, run through the following steps:

1. **Multiply Z^* times s.**
2. **Divide by the desired margin of error, MOE.**
3. **Square it.**
4. **Round *any* fractional amount up to the nearest integer (so you achieve your desired MOE or better).**

For example, suppose that you have to estimate the average exam score for all sixth graders in a district, and you want to be within five points with 95 percent confidence. Suppose that the standard deviation from a pilot study is 12.5 points. The MOE is 5, s is 12.5, and Z^* is 1.96:

1. **Take $1.96 * 12.5 = 24.5$.**
2. **Divide this result by the desired margin of error, which is 5. This gives you $24.5 \div 5 = 4.9$.**
3. **Square your result to get 24.01.**

   You round up any fractional amount, so the sample size must be 25 or more to estimate the average exam score within 5 points.

TIP

No matter what your final answer is, you always round to the next integer when calculating n. Suppose that your calculations say you need $n = 219.16$ people to achieve your desired margin of error. If you round down to 219, the margin of error is larger than what you want. Plus, if the formula says you need 219.16 people, needing a 16th of a person really means you need the whole person.

Suppose that your study has a margin of error of 0.03, and you use 1,000 people. How many people do you need to cut this margin of error in half? You may think the answer is $2 * 1,000 = 2,000$, but you shouldn't jump to conclusions. The answer is $1,000 * 4$, because the formula for n has margin of error squared in the denominator, and the square root of 4 is 2. So if you double the margin of error, you cut the sample size down by four. More importantly, to cut the margin of error in half, you need four times the sample size.

REMEMBER

When you have to find the sample size you need to estimate the population mean, you need the desired margin of error (given in the problem), the level of confidence (given in the problem), and some estimate of the standard deviation based on a previous study. Always round up when you determine sample size. And to cut the margin of error in half, you must quadruple the sample size.

Test Statistic for the Mean

When conducting a hypothesis test for the population mean, you take the sample mean and find out how far it is from the claimed value in terms of a standard score. The standard score is called the *test statistic* (refer to Chapter 13). The formula for the test statistic for the mean is $\frac{(\bar{x} - \mu_o)}{\frac{s}{\sqrt{n}}}$, where μ_o is the claimed value for the *population mean* (the value that sits in the null hypothesis). Here we assume the population standard deviation, σ, is unknown, so we use the sample standard deviation, s. If you do know the population standard deviation, use it instead.

To calculate the test statistic for the sample mean for samples of size 30 or more, follow these steps:

1. **Calculate the sample mean, $\bar{x}$, and the sample standard deviation, s.**
2. **Calculate $\frac{s}{\sqrt{n}}$. Save your answer.**
3. **Take $\bar{x}$ minus μ_o.**
4. **Divide by your result from Step 2.**

For example, suppose a university claims that the average amount of money spent on a football ticket for a certain top-10 football team is \$50. You take a random sample of 100 people and find the average price is \$55 per ticket with a standard deviation of \$15. Do you have enough evidence to say the university's claim is false? The test statistic is $55 - 50$ (which is 5) divided by the quantity 15 over the square root of 100 (which is 1.5), which comes out to be 3.33. This value is out of the range of what you typically expect of a true claim (that is, it's beyond the range −1.96 to 1.96 for what's called a "two-sided hypothesis test"; see Chapter 13). You have enough evidence to say the claim is false.

If you have a larger sample size (say, 200) with the same mean and standard deviation as in the previous example, how would the test statistic compare? It would be larger, because the standard error (denominator of the test statistic) goes down, which makes the test statistic increase.

REMEMBER

The test statistic for testing the population mean is the standardized version of the sample mean. The test statistic is a standard score, so it likely falls between +3 and −3, if you have a large sample size. The further the test statistic is away from 0, the more evidence you have against the null hypothesis. Don't forget to divide by the square root of n in the denominator of the test statistic, because the problem asks about the average, not the individuals. Larger sample sizes drive test statistics out further, so be on the lookout.

Margin of Error for the Sample Proportion

The *margin of error for the sample proportion* (refer to Chapter 10), $\hat{p}$, is the amount by which you expect your sample proportion to vary from sample to sample. The formula for the margin of error for $\hat{p}$, dealing with large enough sample sizes, is $\pm Z^* * \sqrt{\frac{\hat{p}(1-\hat{p})}{n}}$. (Sample sizes are typically large enough with data of this kind, because they often come from surveys of thousands of people or more.)

To calculate the margin of error for $\hat{p}$, follow these steps:

1. **Determine the confidence level and find the appropriate Z^*.**
2. **Calculate the sample proportion, $\hat{p}$, and the sample size, n.**

   The sample proportion is the number of individuals in the desired category divided by the sample size.
3. **Multiply $\hat{p}$ times (1 minus $\hat{p}$).**
4. **Divide by n.**
5. **Take the square root.**
6. **Multiply by your answer from Step 1.**

For example, suppose that you want to estimate the proportion of all adults in the United States who floss their teeth daily. You take a random sample of 1,000 people and find that 650 report flossing daily. What's your estimate of the proportion of all adults in the United States that floss daily?

1. **Suppose that you want to be 95 percent confident in your results.**

   That means $Z^* = 1.96$.
2. **Find the sample proportion, which in this case is $650 \div 1{,}000 = 0.65$.**
3. **Next, you find $0.65(1 - 0.65) = 0.2275$.**
4. **Now, take that result and divide by 1,000 to get 0.00023.**

5. **Take the square root, which is 0.015.**
6. **(Don't forget!) Multiply by your Z^* value from Step 1 (which is 1.96).**

   This gives you 0.0294, or 2.94 percent. You now have the margin of error.

If the sample size gets larger, the denominator of the fraction gets larger, which makes the entire margin of error smaller. However, wanting more confidence in the results creates a larger Z^* value, which makes the margin of error go up. Also, different values of the population proportion, $\hat{p}$, result in larger variances (and hence in a larger margin of error). The value of $\hat{p}$ that gives the largest margin of error is $\hat{p} = \frac{1}{2}$, and this happens when the population is split 50-50, so the results have the most variability from person to person.

REMEMBER

The margin of error is plus or minus (±) the amount calculated in the previous steps, and the part that follows the ± sign. More confidence and/or more variability in the population means the margin of error increases; larger samples decrease the margin of error.

Sample Size Needed for Estimating *p*

If you want a confidence interval for the population mean with a certain margin of error, you can figure out the sample size needed before you collect any data. If you come up with a value of $\hat{p}$ with a pilot study first, the sample size for estimating the proportion *p* is $n = \frac{(Z^*)^2 * \hat{p} * (1-\hat{p})}{\text{MOE}^2}$, where Z^* is the standard normal value for your confidence level, and $\hat{p}$ is the proportion who fall into the category of interest from the pilot study.

To calculate the sample size for the proportion, run through the following steps:

1. **Multiply Z^* squared times $\hat{p}$ times $(1-\hat{p})$.**
2. **Divide by the MOE squared (MOE is the desired margin of error).**
3. **You should round *any* fractional amount up to the nearest integer (so you achieve the desired MOE or better).**

For example, suppose that you want to estimate the proportion of people in the United States in favor of Issue X to within 2.5 percentage points, with 95 percent confidence. A small pilot study finds that 58 percent are in favor of the issue:

1. **Take 1.96 squared (for 95 percent confidence) times 0.58 times $(1-0.58)$, which gives you $3.8416 * 0.58 * 0.42 = 0.9358$.**

   The MOE is 0.025 (squared is 0.000625).

2. **Now take $0.9358 \div 0.000625 = 1{,}497.28$.**
3. **Round up any fractional amount, so 1,497.28 rounds up to 1,498.**

   You need a sample size of at least 1,498 people to achieve a margin of error of 2.5 percent or less.

If you can't conduct a pilot study, you can substitute ½ for the sample proportion to get a conservative estimate for the sample size. You use this method in a worst-case scenario to get a conservative answer for *n*.

The formula for the sample size in the case where you use $p = ½$ is $\frac{(Z^*)^2}{(4)(\text{MOE})^2}$. To calculate the sample size:

1. **Calculate 4 times the square of the MOE. Save your answer.**
2. **Take Z^* squared and divide the answer by the result from Step 1.**

In the previous example without the pilot study result, take 4 times 0.025 squared $= 4 * 0.000625 = 0.0025$. Now you take 1.96 squared (3.8416) and divide by 0.0025 to get 1,536.64, and round up to 1,537. You need a few more people than you had with the pilot study, because $p = ½$ is the situation where the population has the most variability and where you need more data.

REMEMBER

When you have to find the sample size needed to estimate the population proportion, you need the desired margin of error (given in the problem), the level of confidence (given in the problem), and some estimate of *p* based on a previous study. If you don't have any previous information, use $p = ½$ as a worst-case scenario. Always round up when you determine sample size.

Test Statistic for the Proportion

When conducting a hypothesis test for the population mean, you take the sample mean and find out how far it is from the claimed value in terms of a standard score. The standard score is called the *test statistic* (see Chapter 13). The formula for the test statistic for the population proportion is $\frac{\hat{p} - p_o}{\sqrt{\frac{p_o(1-p_o)}{n}}}$, where p_o is the claimed value for the population proportion (the number sitting in the null hypothesis).

To calculate the test statistic for the proportion, follow these steps:

1. **Calculate the sample proportion, $\hat{p}$.**

   The sample proportion is the number of individuals in the desired category divided by the sample size.

2. **Calculate p_o times $(1 - p_o)$ divided by *n*.**
3. **Take the square root. Save your answer.**
4. **Take $\hat{p}$ minus p_o.**
5. **Divide the result by your answer from Step 3.**

Suppose you hear a claim that only 60 percent of people in the United States floss their teeth daily, and you suspect the percentage is higher than that. You take a random sample of 1,000 people and find that 650 report flossing daily. Do you have enough evidence to say the claim is wrong? You find out by conducting a hypothesis test for p, the proportion of all people who floss daily. Ho is $p = 0.60$ and Ha is $p > 0.60$. Your sample gives you $\hat{p} = 650 \div 1{,}000 = 0.65$. The test statistic is $\dfrac{\hat{p} - p_o}{\sqrt{\dfrac{p_o(1 - p_o)}{n}}} = \dfrac{0.65 - 0.60}{\sqrt{\dfrac{0.60(1 - 0.60)}{1{,}000}}} = \dfrac{0.05}{\sqrt{0.00024}} = 3.23$. This test statistic is large (beyond a Z-value of 1.64), so you deny the claim.

REMEMBER

The test statistic for testing the population proportion is the standardized version of the sample proportion. To find the sample proportion, you take the number of "successes" (people who have the characteristic of interest) and divide by the total sample size (n). This should give you a decimal between 0 and 1. The test statistic is a standard score, so it should fall between +3 and –3, if you have a large sample size. The further the test statistic is away from 0, the more evidence you have against the null hypothesis.

See Chapter 13 for more on test statistics in general, including those for one or two means and one or two proportions.

» Being on the lookout for misleading statistics

Chapter 21

Ten Ways to Spot Common Statistical Mistakes

You shouldn't accept every statistic without question. When you know what questions to ask and what to look for, spotting problems in statistics and feeling empowered to decide whether to believe the numbers you see becomes much easier. It also helps to be as specific as possible in terms of critiquing a statistic. Saying exactly why a statistic is incorrect or misleading is much more effective than just saying that a statistic is "bogus" or "biased."

In this chapter, I summarize the big ideas for spotting errors or misleading statistics, show you how to report them with the right lingo, and give tips for avoiding them altogether when you do your own statistical analyses.

Scrutinizing Graphs

Graphs can be misleading in a number of ways: by what they show and by what they leave out. Here are some tips for looking at graphs and spotting misleading info:

» **The values on certain types of graphs should sum to 1.** For example, a pie chart or a relative frequency histogram/bar chart. In general, for any chart/graph where the group is broken down into categories that do not overlap, and percents are shown, those percents should sum to 1. (See Chapter 3 for more on graphs using percents.)

- **Watch for distortions that come with graphs that use pictures to represent amounts.** For example, a picture twice as wide and twice as high as another is actually four times larger in terms of actual area.
- **Any graph that shows percents should also show the total number of individuals to give you some idea of the precision of the results.** For example, just saying 80 percent of the people support Issue X and 20 percent oppose it without telling how many were surveyed is misleading. (See Chapter 3 for sample size and graph info.)
- **Watch for a distortion of the scale.** This is one of the biggest problems with misleading graphs. Creators do it by making the increments larger on the *y*-axis than they should be (which makes differences seem smaller than they should) or by making the increments smaller on the *y*-axis than they should be (which makes differences seem larger than they should). (See Chapter 4 for more on scale distortion of graphs.)
- **Line Graphs (time charts) that distort the time (*x*-axis) scale make the data look overly varied and volatile (if time increments are too small) or overly smooth (if time increments are too large).** Time increments should also have an equal distance between them. For example, a time axis with the years 1970, 1980, and 1985 all equally spaced is misleading. (See Chapter 4 for more on line graphs.)
- **Make sure that graphs showing money or similar units over time are adjusted for inflation over time as well.** For example, if you want to show the change in the average cost of a home over a 50-year period, you need to adjust those figures for inflation or the differences will look bigger than they really are. (See Chapter 4 for more on scale and line graphs.)
- **When comparing two or more histograms, make sure that the *x*- and *y*-axes are the same.** For example, if you want to compare the lengths of fish in two ponds, and one pond is much more variable than the other, showing them on the same scale helps you spot the difference in variability. Showing them on different scales clouds that difference. (See Chapter 4 for more on histograms and scale.)

Searching for and Specifying Bias

Bias in statistics refers to a systematic error that either consistently overestimates or consistently underestimates the true value. Bias can happen in many different ways and places, each pointing to errors in the statistical process:

- **During the sampling process, if the sample is selected in a way that favors certain groups over others:** For example, an Internet survey favors people who are online more often and who frequent certain sites. (Chapter 15 talks more about errors that can happen during sampling.)
- **During the data collection process:** Bias can happen in the wording of the questions, the way the interviews are held, or when the subjects feel they can't be honest about their answers. For example, "Have you ever cheated on an exam?" (See Chapter 15 for more ways that the data collection process can go bad.)

» **In the recording of the data:** For example, using a weighing scale that's 5 pounds off in the high direction. (See Chapters 15 and 16 for more errors to avoid in data recording.)

» **During experiments:** If the researchers know which subjects get the real treatment and which get the placebo, they might pay more attention to those on the real treatment and encourage them to report a positive outcome. (If you want more examples of how experiments can be biased, see Chapter 16.)

» **During the data analysis:** If a researcher throws out data that don't fit his or her model, the data will be biased toward the results the researcher wanted. For example, suppose that all the patients in her sample reacted the "right way" except one who had a bad reaction. She might be tempted to encourage that person to drop out of the study and to not use his data. This creates a biased overall result because it doesn't include the one person who reacted badly, and that person might represent a larger segment of the population. Leaving data out of a study is illegal, of course, and doesn't happen often, but it does happen. (Chapter 16 helps you look more critically at the way experiments are designed.)

Marking the Margin of Error

Sampling errors occur every time someone tries to estimate a population value with a sample, because sample results vary from sample to sample. To evaluate statistical results, you need a measure of their accuracy. You typically measure the accuracy through the *margin of error*. The margin of error tells you how much the researcher expects his or her results to vary from random sample to random sample.

When researchers or the media members fail to report the margin of error, they leave it up to the consumer to wonder about the accuracy of the results; or worse, they just assume that all the info is fine, when in many cases it isn't! Survey results shown on television, until quite recently, rarely include a margin of error. Still, many newspaper, magazine, and Internet surveys fail to report a margin of error, or they report a meaningless margin of error caused by biased data. (For more practice with margin of error, see Chapter 10.)

REMEMBER

When you look at statistical results that point to an estimated number (for example, the percentage of people who think the president is doing a good job), always check for the margin of error. If it isn't included, ask for it! (Or if given enough other pertinent information, you can calculate the margin of error yourself, using the formulas in Chapters 10 and 20.)

Scanning for Sample Size

The quantity of information is always important in terms of assessing the accuracy of a statistic. The more information that goes into a statistic, the more accurate the statistic will be — as long as the information isn't biased. The consumer of the statistical information needs to be able to assess the accuracy of the information, and for that, you need to look at how researchers collected the information and how much they collected (the sample size).

Many charts and graphs that appear in the media don't include a sample size. You may also realize that many headlines aren't exactly what they seem to be when you see that the details of the articles reveal either small sample sizes (reducing reliability in the results) or no information at all about the sample sizes. (Remember the gum ad that says, "Four out of five dentists surveyed choose this gum for their patients who chew gum?" What if they really did ask only five dentists?) Chapter 10 deals more in-depth with the whole sample-size issue and how it affects margin of error.

How large is "large enough"? A survey of 2,500 people has a margin of error of only about 2 percentage points. For experiments and other types of studies asking for numerical information, most statisticians say you need at least 30 people in each subgroup in the study to have accurate data.

Always look for the sample size before you make decisions about statistical information. The smaller the sample size, the less reliable the information. If the sample size is missing from the article, get a copy of the full report of the study, contact the researcher, or contact the journalist who wrote the article.

Studying Sample Selection (Gotta Be Random)

From survey data to results of medical studies, most statistics are based on data collected from samples of individuals rather than from entire populations because of cost and time considerations. Plus, you don't need a huge sample to be amazingly accurate — *if* the sample you study is representative of the population. For example, a survey of 2,500 people has a margin of error of roughly plus or minus only 2 percent. How can you ensure that the sample represents the population? The best way is to *randomly* select the individuals from the population. A random sample is a subset of the population selected in such a way that each sample of the same size in the population has an equal chance of being selected (like drawing names out of a hat). No systematic favoritism or exclusion is involved in a random sample.

Unfortunately, many surveys and studies aren't based on random samples of individuals. For example, many medical studies involve volunteers who aren't randomly selected. It isn't ethical or practical to phone people and say, "We chose you at random to participate in our sleeping study. You need to come down to our lab and stay for four nights." The best you can do here is study the volunteers to see how closely they represent the population and report that, or ask for certain types of volunteers.

Polls and surveys should also be based on randomly selected individuals — a much easier task compared to medical studies. However, many surveys aren't based on random samples. For example, television polls asking viewers to "call us with your opinion" aren't based on random samples. These surveys don't give the entire population an equal chance of selection; in fact, in these examples, the people choose themselves. (Chapter 15 looks at sampling methods and how to scrutinize them.)

Before you make any decisions about statistical results from a survey or a study, look at the selection process. If the sample participants aren't selected randomly, take the results with a grain of salt.

Checking for Confounding Variables

A *confounding variable* is an ignored variable that influences the results of the study, creating a confusing, or confounding, effect. For example, suppose that a researcher tries to say that eating seaweed helps you live longer. When you examine the study further, you find out that the researcher based it on a sample of people who regularly eat seaweed in their diets and are over the age of 100. Did the seaweed cause them to live longer, or did something else contribute to their longevity? You can't tell because of the confounding variables — including exercise, water consumption, diet, and sleeping patterns — that can also cause longer life.

A common error in research studies is ignoring confounding variables, leaving the results open to scrutiny. The best way to control for confounding variables is to do a *designed experiment*. You compare the results from two groups, and any significant differences are attributable to the treatment and nothing else (in an ideal world). The seaweed study isn't a designed experiment; it's an *observational study*. In observational studies, you exercise no control over any variables; you merely observe people and record information. (If you want to investigate confounding variables further, see Chapter 16.)

Observational studies are great for surveys and polls, but they're great for showing cause-and-effect relationships because they don't control for confounding variables. A designed experiment provides much stronger evidence for cause-and-effect relationships and should be done whenever ethically possible.

Considering Correlation

Correlation is one of the most misunderstood and misused statistical terms exercised by researchers, the media, and the general public. (All you ever wanted to know about correlation is located in Chapter 18.) I can think of three typical mistakes regarding correlation:

- **Applying correlation to non-quantitative variables:** Correlation applies only to two numerical (quantitative) variables; for example, height and weight. Correlation *doesn't* apply to categorical variables such as political party and gender. So if you hear someone say, "It appears that the voting pattern is correlated with gender . . .," the speaker is incorrect. Voting pattern and gender may be associated, but they can't be correlated, according to the statistical definition of correlation.
- **Not realizing what correlation does and doesn't measure:** A correlation measures the strength and direction of the *linear* relationship between two numerical variables. If two variables aren't correlated, it doesn't mean they have no relationship at all, however. For example, bacteria multiply at an *exponential* rate over time (their numbers explode, doubling faster and faster), not at a *linear* rate (which is a steady increase over time).
- **Thinking that correlation automatically means cause and effect:** For example, suppose someone reports that more people who drink diet soda have brain tumors than people who don't. Don't panic just yet. The researcher may have discovered a freak occurrence in nature. At most, it means more research needs to be done (beyond observation) to show that diet soda *causes* brain tumors.

Correlation applies only to numerical variables, and it implies only that the variables are related in a linear way (up or down). Correlation doesn't necessarily mean cause and effect; you see this case only if other possible explanations can be ruled out and/or controlled for.

Doing the Math

Just because someone in the media reports a statistic doesn't make it correct. In fact, errors appear all the time (by error or by design), and you need to be on the lookout for them. Here are some tips for spotting botched numbers:

- **Check to be sure that all the figures add up to the reported value.** With pie charts, be sure all the percentages add up to 100 percent.
- **Double-check even the most basic of calculations.** For example, a chart says 83 percent of Americans are in favor of an issue, but the report says seven out of every eight Americans are in favor of the issue. Are these figures the same? (No, $7 \div 8 = 87.5$ percent; five out of six is about 83 percent.)
- **Look for the *response rate* of a survey; don't just be happy with the number of participants.** The *response rate* is the number of people who respond divided by the total number of people surveyed. If the response rate is much lower than 70 percent, the results may be biased, because you don't know what the nonrespondents would say.
- **Question the type of statistic used; is it appropriate?** For example, the number of crimes went up, but so did the population size. Researchers should report the *crime rate* (number of crimes per capita).

For more examples on how graphs and statistics can go wrong, see Chapters 3 and 4 (graphs) and Chapters 1 and 2 (statistics). If you want to brush up on your math and avoid some common math mistakes, take a look at Chapter 19.

Statistics is based on formulas and calculations that don't know any better; the people plugging in the numbers, however, do (or should) know better, but they may not (or they don't want you to catch on). Consumers of information (certified skeptics) have to take action and get to the bottom of it. The best policy is to ask questions!

Detecting Selective Reporting

Another bad scenario plays out when a researcher reports his or her one *"statistically significant"* result (one unlikely to occur simply by chance) but leaves out the fact that he or she actually conducted hundreds of other tests, each of them nonsignificant. If you knew about all the other tests, you would wonder whether the significant result is really meaningful or whether you can attribute it to chance, simply because the researcher did so many tests. These bad practices are what statisticians like to call "data snooping" or "data fishing" (see Chapter 14).

How can you protect yourself against misleading results due to data fishing? Find out more details about the study, such as how many tests were done, how many insignificant results came out, and what results appear significant. In other words, get the whole story if you can so that you can put the significant results into perspective.

To spot fudged numbers and errors of omission, the best policy is to remember that if something looks too good to be true, it probably is. Don't just go on the first result that you hear, especially if it makes big news. Wait to see whether others can verify and replicate the result.

Avoiding the Anecdote

An *anecdote* is a story based on a single person's experience or situation. A few examples: the waitress who wins the lottery — twice; the cat that learns how to ride a bicycle; the woman who loses 100 pounds in two days on the miracle potato diet; or the celebrity who uses the over-the-counter hair color that you see on television (yeah, right). Anecdotes make great news — the more sensational, the better. But sensational stories are outliers from the norm of life. They don't happen to most of us.

An anecdote is a data set with a sample size of only one. You have no information to compare the story to, no statistics to analyze, no possible explanations or information to go on; you have only a single story. Don't let anecdotes have much influence over you. You should rely on scientific studies and statistical information based on large random samples of individuals who represent their entire populations (not just a single situation).

The best way to respond when someone tries to persuade you by telling you an anecdote is to say: "Show me the data!"

Appendix

Tables for Reference

This appendix includes tables for finding probabilities and/or critical values for the three distributions used in this book: the Z-distribution (standard normal), the t-distribution, and the binomial distribution. I explain how to use each table since they are all a bit different.

Z-Table

The Z-table shows less-than-or-equal-to probabilities for the Z-distribution; that is, $P(Z \leq z)$ for a given z value. (See Chapter 6 for more on the Z-distribution.) To use this table, do the following:

1. **Determine the z-value for your particular problem.**

   The z-value should have one leading digit before the decimal point (positive, negative, or zero) and two digits after the decimal point; for example $z = 128$, -2.69, or 0.13.

2. **Find the row of the table corresponding to the leading digit and first digit after the decimal point.**

   For example, if your z-value is 1.28, look in the "1.2" row; if $z = -1.28$, look in the "−1.2" row.

3. **Find the column corresponding to the second digit after the decimal point.**

   For example, if your z-value is 1.28 or −1.28, look in the ".08" column.

4. **Intersect the row and column from Steps 2 and 3.**

   This number is the probability that Z is less than or equal to your z-value. In other words, you've found $P(Z \leq z)$. For example, if $z = 1.28$, you see $P(Z \leq 1.28) = 0.8997$. For $z = -1.28$, you see $P(Z \leq -1.28) = 0.1003$.

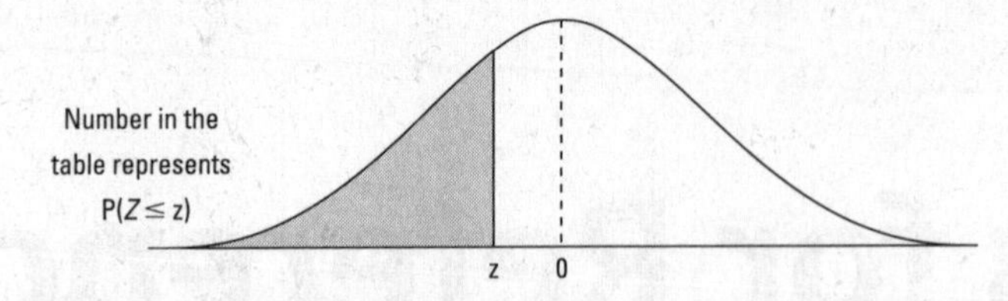

z	0.00	0.01	0.02	0.03	0.04	0.05	0.06	0.07	0.08	0.09
−3.6	.0002	.0002	.0001	.0001	.0001	.0001	.0001	.0001	.0001	.0001
−3.5	.0002	.0002	.0002	.0002	.0002	.0002	.0002	.0002	.0002	.0002
−3.4	.0003	.0003	.0003	.0003	.0003	.0003	.0003	.0003	.0002	.0002
−3.3	.0005	.0005	.0005	.0004	.0004	.0004	.0004	.0004	.0003	.0003
−3.2	.0007	.0007	.0006	.0006	.0006	.0006	.0006	.0005	.0005	.0005
−3.1	.0010	.0009	.0009	.0009	.0008	.0008	.0008	.0008	.0007	.0007
−3.0	.0013	.0013	.0013	.0012	.0012	.0011	.0011	.0011	.0010	.0010
−2.9	.0019	.0018	.0018	.0017	.0016	.0016	.0015	.0015	.0014	.0014
−2.8	.0026	.0025	.0024	.0023	.0023	.0022	.0021	.0021	.0020	.0019
−2.7	.0035	.0034	.0033	.0032	.0031	.0030	.0029	.0028	.0027	.0026
−2.6	.0047	.0045	.0044	.0043	.0041	.0040	.0039	.0038	.0037	.0036
−2.5	.0062	.0060	.0059	.0057	.0055	.0054	.0052	.0051	.0049	.0048
−2.4	.0082	.0080	.0078	.0075	.0073	.0071	.0069	.0068	.0066	.0064
−2.3	.0107	.0104	.0102	.0099	.0096	.0094	.0091	.0089	.0087	.0084
−2.2	.0139	.0136	.0132	.0129	.0125	.0122	.0119	.0116	.0113	.0110
−2.1	.0179	.0174	.0170	.0166	.0162	.0158	.0154	.0150	.0146	.0143
−2.0	.0228	.0222	.0217	.0212	.0207	.0202	.0197	.0192	.0188	.0183
−1.9	.0287	.0281	.0274	.0268	.0262	.0256	.0250	.0244	.0239	.0233
−1.8	.0359	.0351	.0344	.0336	.0329	.0322	.0314	.0307	.0301	.0294
−1.7	.0446	.0436	.0427	.0418	.0409	.0401	.0392	.0384	.0375	.0367
−1.6	.0548	.0537	.0526	.0516	.0505	.0495	.0485	.0475	.0465	.0455
−1.5	.0668	.0655	.0643	.0630	.0618	.0606	.0594	.0582	.0571	.0559
−1.4	.0808	.0793	.0778	.0764	.0749	.0735	.0721	.0708	.0694	.0681
−1.3	.0968	.0951	.0934	.0918	.0901	.0885	.0869	.0853	.0838	.0823
−1.2	.1151	.1131	.1112	.1093	.1075	.1056	.1038	.1020	.1003	.0985
−1.1	.1357	.1335	.1314	.1292	.1271	.1251	.1230	.1210	.1190	.1170
−1.0	.1587	.1562	.1539	.1515	.1492	.1469	.1446	.1423	.1401	.1379
−0.9	.1841	.1814	.1788	.1762	.1736	.1711	.1685	.1660	.1635	.1611
−0.8	.2119	.2090	.2061	.2033	.2005	.1977	.1949	.1922	.1894	.1867
−0.7	.2420	.2389	.2358	.2327	.2296	.2266	.2236	.2206	.2177	.2148
−0.6	.2743	.2709	.2676	.2643	.2611	.2578	.2546	.2514	.2483	.2451
−0.5	.3085	.3050	.3015	.2981	.2946	.2912	.2877	.2843	.2810	.2776
−0.4	.3446	.3409	.3372	.3336	.3300	.3264	.3228	.3192	.3156	.3121
−0.3	.3821	.3783	.3745	.3707	.3669	.3632	.3594	.3557	.3520	.3483
−0.2	.4207	.4168	.4129	.4090	.4052	.4013	.3974	.3936	.3897	.3859
−0.1	.4602	.4562	.4522	.4483	.4443	.4404	.4364	.4325	.4286	.4247
−0.0	.5000	.4960	.4920	.4880	.4840	.4801	.4761	.4721	.4681	.4641

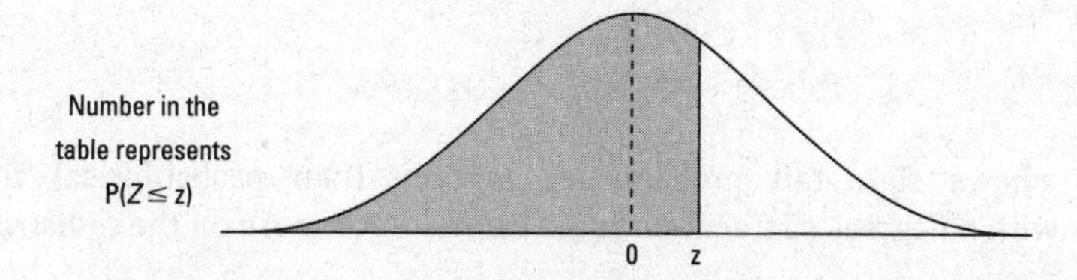

z	0.00	0.01	0.02	0.03	0.04	0.05	0.06	0.07	0.08	0.09
0.0	.5000	.5040	.5080	.5120	.5160	.5199	.5239	.5279	.5319	.5359
0.1	.5398	.5438	.5478	.5517	.5557	.5596	.5636	.5675	.5714	.5753
0.2	.5793	.5832	.5871	.5910	.5948	.5987	.6026	.6064	.6103	.6141
0.3	.6179	.6217	.6255	.6293	.6331	.6368	.6406	.6443	.6480	.6517
0.4	.6554	.6591	.6628	.6664	.6700	.6736	.6772	.6808	.6844	.6879
0.5	.6915	.6950	.6985	.7019	.7054	.7088	.7123	.7157	.7190	.7224
0.6	.7257	.7291	.7324	.7357	.7389	.7422	.7454	.7486	.7517	.7549
0.7	.7580	.7611	.7642	.7673	.7704	.7734	.7764	.7794	.7823	.7852
0.8	.7881	.7910	.7939	.7967	.7995	.8023	.8051	.8078	.8106	.8133
0.9	.8159	.8186	.8212	.8238	.8264	.8289	.8315	.8340	.8365	.8389
1.0	.8413	.8438	.8461	.8485	.8508	.8531	.8554	.8577	.8599	.8621
1.1	.8643	.8665	.8686	.8708	.8729	.8749	.8770	.8790	.8810	.8830
1.2	.8849	.8869	.8888	.8907	.8925	.8944	.8962	.8980	.8997	.9015
1.3	.9032	.9049	.9066	.9082	.9099	.9115	.9131	.9147	.9162	.9177
1.4	.9192	.9207	.9222	.9236	.9251	.9265	.9279	.9292	.9306	.9319
1.5	.9332	.9345	.9357	.9370	.9382	.9394	.9406	.9418	.9429	.9441
1.6	.9452	.9463	.9474	.9484	.9495	.9505	.9515	.9525	.9535	.9545
1.7	.9554	.9564	.9573	.9582	.9591	.9599	.9608	.9616	.9625	.9633
1.8	.9641	.9649	.9656	.9664	.9671	.9678	.9686	.9693	.9699	.9706
1.9	.9713	.9719	.9726	.9732	.9738	.9744	.9750	.9756	.9761	.9767
2.0	.9772	.9778	.9783	.9788	.9793	.9798	.9803	.9808	.9812	.9817
2.1	.9821	.9826	.9830	.9834	.9838	.9842	.9846	.9850	.9854	.9857
2.2	.9861	.9864	.9868	.9871	.9875	.9878	.9881	.9884	.9887	.9890
2.3	.9893	.9896	.9898	.9901	.9904	.9906	.9909	.9911	.9913	.9916
2.4	.9918	.9920	.9922	.9925	.9927	.9929	.9931	.9932	.9934	.9936
2.5	.9938	.9940	.9941	.9943	.9945	.9946	.9948	.9949	.9951	.9952
2.6	.9953	.9955	.9956	.9957	.9959	.9960	.9961	.9962	.9963	.9964
2.7	.9965	.9966	.9967	.9968	.9969	.9970	.9971	.9972	.9973	.9974
2.8	.9974	.9975	.9976	.9977	.9977	.9978	.9979	.9979	.9980	.9981
2.9	.9981	.9982	.9982	.9983	.9984	.9984	.9985	.9985	.9986	.9986
3.0	.9987	.9987	.9987	.9988	.9988	.9989	.9989	.9989	.9990	.9990
3.1	.9990	.9991	.9991	.9991	.9992	.9992	.9992	.9992	.9993	.9993
3.2	.9993	.9993	.9994	.9994	.9994	.9994	.9994	.9995	.9995	.9995
3.3	.9995	.9995	.9995	.9996	.9996	.9996	.9996	.9996	.9996	.9997
3.4	.9997	.9997	.9997	.9997	.9997	.9997	.9997	.9997	.9997	.9998
3.5	.9998	.9998	.9998	.9998	.9998	.9998	.9998	.9998	.9998	.9998
3.6	.9998	.9998	.9999	.9999	.9999	.9999	.9999	.9999	.9999	.9999

t-Table

The t-table shows right-tail probabilities (greater-than probabilities) for selected t-distributions with n degrees of freedom. (See Chapter 8 for more on the t-distribution.)

Follow these steps to use the t-table to find greater-than probabilities and p-values for hypothesis involving t (see Chapter 13):

1. **Find the t-value for which you want the greater-than probability (call it t), and find the sample size (for example, n).**
2. **Find the row corresponding to the degrees of freedom (df) for your problem (for example, $n - 1$). Go across that row to find the two t-values between which your t-falls.**

   For example, if your t is 1.60 and your n is 7, you look in the row for df $= 7 - 1 = 6$. Across that row you find your t lies between t-values 1.44 and 1.94.
3. **Go to the top of the columns containing the two t-values from Step 2.**

   The greater-than probability for your t-value is between the two values at the top of these columns. For example, if your df $= 6$, your $t = 1.60$ is between t-values 1.44 and 1.94, so the greater-than probability for your t is between 0.10 (column heading for $t = 1.44$); and 0.05 (column heading for $t = 1.94$).

TIP

The row near the bottom with the Z in the df column gives greater-than probabilities from the Z-distribution (Chapter 8 shows the relationship between t and Z).

You can also use the t-table to find critical values (t^*) for a confidence interval involving t (see Chapter 11):

1. **Determine the confidence level you need (as a percentage).**
2. **Determine the sample size (for example, n).**
3. **Look at the bottom row of the table where the percentages are shown. Find your % confidence level there.**
4. **Intersect this column with the row representing your degrees of freedom (df). This is the t-value you need for your confidence interval.**

   For example, a 95% confidence interval with df $= 6$ has $t^* = 2.45$. (Find 95% on the last line and go up to row 6.)

Numbers in each row of the table are values on a *t*-distribution with (*df*) degrees of freedom for selected right-tail (greater-than) probabilities (*p*).

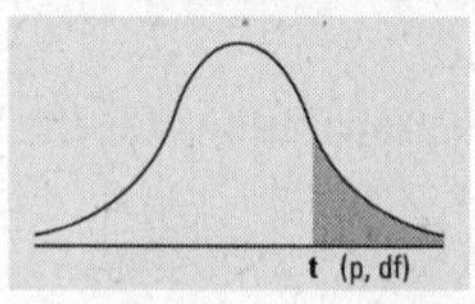

df/p	0.40	0.25	0.10	0.05	0.025	0.01	0.005	0.0005
1	0.324920	1.000000	3.077684	6.313752	12.70620	31.82052	63.65674	636.6192
2	0.288675	0.816497	1.885618	2.919986	4.30265	6.96456	9.92484	31.5991
3	0.276671	0.764892	1.637744	2.353363	3.18245	4.54070	5.84091	12.9240
4	0270722	0.740697	1.533206	2.131847	2.77645	3.74695	4.60409	8.6103
5	0.267181	0.726687	1.475884	2.015048	2.57058	3.36493	4.03214	6.8688
6	0.264835	0.717558	1.439756	1.943180	2.44691	3.14267	3.70743	5.9588
7	0.263167	0.711142	1.414924	1.894579	2.36462	2.99795	3.49948	5.4079
8	0.261921	0.706387	1.396815	1.859548	2.30600	2.89646	3.35539	5.0413
9	0.260955	0.702722	1.383029	1.833113	2.26216	2.82144	3.24984	4.7809
10	0260185	0.699812	1.372184	1.812461	2.22814	2.76377	3.16927	4.5869
11	0259556	0.697445	1.363430	1.795885	2.20099	2.71808	3.10581	4.4370
12	0259033	0.695483	1.356217	1.782288	2.17881	2.68100	3.05454	43178
13	0.258591	0.693829	1.350171	1.770933	2.16037	2.65031	3.01228	4.2208
14	0.258213	0.692417	1.345030	1.761310	2.14479	2.62449	2.97684	4.1405
15	0.257885	0.691197	1.340606	1.753050	2.13145	2.60248	2.94671	4.0728
16	0257599	0.690132	1.336757	1.745884	2.11991	2.58349	2.92078	4.0150
17	0.257347	0.689195	1.333379	1.739607	2.10982	2.56693	2.89823	3.9651
18	0.257123	0.688364	1.330391	1.734064	2.10092	2.55238	2.87844	3.9216
19	0.256923	0.687621	1.327728	1.729133	2.09302	2.53948	2.86093	3.8834
20	0.256743	0.686954	1.325341	1.724718	2.08596	2.52798	2.84534	3.8495
21	0.256580	0.686352	1.323188	1.720743	2.07961	2.51765	2.83136	3.8193
22	0256432	0.685805	1.321237	1.717144	2.07387	2.50832	2.81876	3.7921
23	0256297	0.685306	1.319460	1.713872	2.06866	2.49987	2.80734	3.7676
24	0.256173	0.684850	1.317836	1.710882	2.06390	2.49216	2.79694	3.7454
25	0.256060	0.684430	1.316345	1.708141	2.05954	2.48511	2.78744	3.7251
26	0.255955	0.684043	1.314972	1.705618	2.05553	2.47863	2.77871	3.7066
27	0.255858	0.683685	1.313703	1.703288	2.05183	2.47266	2.77068	3.6896
28	0.255768	0.683353	1.312527	1.701131	2.04841	2.46714	2.76326	3.6739
29	0.255684	0.683044	1.311434	1.699127	2.04523	2.46202	2.75639	3.6594
30	0.255605	0.682756	1.310415	1.697261	2.04227	2.45726	2.75000	3.6460
z	0.253347	0.674490	1.281552	1.644854	1.95996	2.32635	2.57583	3.2905
CI	———	———	80%	90%	95%	98%	99%	99.9%

Binomial Table

The binomial table shows probabilities for the binomial distribution (see Chapter 7).

To use the binomial table do the following:

1. **Find these three numbers for your particular problem:**
   - The sample size, n
   - The probability of success, p
   - The x-value for which you want $p(X = x)$
2. **Find the section of the binomial table that's devoted to your n.**
3. **Look at the row for your x-value and the column for your p.**
4. **Intersect that row and column. You have found $p(X = x)$.**
5. **To get the probability of being less than, greater than, greater than or equal to, less than or equal to, or between two values of X, you add the appropriate values of the table using the steps found in Chapter 7.**

Binomial probabilities:

$$\binom{n}{x} p^x (1-p)^{n-x}$$

Numbers in the table represent $p(X=x)$ for a binomial distribution with n trials and probability of success p.

		p										
n	***x***	**0.1**	**0.2**	**0.25**	**0.3**	**0.4**	**0.5**	**0.6**	**0.7**	**0.75**	**0.8**	**0.9**
1	*0*	0.900	0.800	0.750	0.700	0.600	0.500	0.400	0.300	0.250	0.200	0.100
	1	0.100	0.200	0.250	0.300	0.400	0.500	0.600	0.700	0.750	0.800	0.900
2	*0*	0.810	0.640	0.563	0.490	0.360	0.250	0.160	0.090	0.063	0.040	0.010
	1	0.180	0.320	0.375	0.420	0.480	0.500	0.480	0.420	0.375	0.320	0.180
	2	0.010	0.040	0.063	0.090	0.160	0.250	0.360	0.490	0.563	0.640	0.810
3	*0*	0.729	0.512	0.422	0.343	0.216	0.125	0.064	0.027	0.016	0.008	0.001
	1	0.243	0.384	0.422	0.441	0.432	0.375	0.288	0.189	0.141	0.096	0.027
	2	0.027	0.096	0.141	0.189	0.288	0.375	0.432	0.441	0.422	0.384	0.243
	3	0.001	0.008	0.016	0.027	0.064	0.125	0.216	0.343	0.422	0.512	0.729
4	*0*	0.656	0.410	0.316	0.240	0.130	0.063	0.026	0.008	0.004	0.002	0.000
	1	0.292	0.410	0.422	0.412	0.346	0.250	0.154	0.076	0.047	0.026	0.004
	2	0.049	0.154	0.211	0.265	0.346	0.375	0.346	0.265	0.211	0.154	0.049
	3	0.004	0.026	0.047	0.076	0.154	0.250	0.346	0.412	0.422	0.410	0.292
	4	0.000	0.002	0.004	0.008	0.026	0.063	0.130	0.240	0.316	0.410	0.656
5	*0*	0.590	0.328	0.237	0.168	0.078	0.031	0.010	0.002	0.001	0.000	0.000
	1	0.328	0.410	0.396	0.360	0.259	0.156	0.077	0.028	0.015	0.006	0.000
	2	0.073	0.205	0.264	0.309	0.346	0.312	0.230	0.132	0.088	0.051	0.008
	3	0.008	0.051	0.088	0.132	0.230	0.312	0.346	0.309	0.264	0.205	0.073
	4	0.000	0.006	0.015	0.028	0.077	0.156	0.259	0.360	0.396	0.410	0.328
	5	0.000	0.000	0.001	0.002	0.010	0.031	0.078	0.168	0.237	0.328	0.590
6	*0*	0.531	0.262	0.178	0.118	0.047	0.016	0.004	0.001	0.000	0.000	0.000
	1	0.354	0.393	0.356	0.303	0.187	0.094	0.037	0.010	0.004	0.002	0.000
	2	0.098	0.246	0.297	0.324	0.311	0.234	0.138	0.060	0.033	0.015	0.001
	3	0.015	0.082	0.132	0.185	0.276	0.313	0.276	0.185	0.132	0.082	0.015
	4	0.001	0.015	0.033	0.060	0.138	0.234	0.311	0.324	0.297	0.246	0.098
	5	0.000	0.002	0.004	0.010	0.037	0.094	0.187	0.303	0.356	0.393	0.354
	6	0.000	0.000	0.000	0.001	0.004	0.016	0.047	0.118	0.178	0.262	0.531
7	*0*	0.478	0.210	0.133	0.082	0.028	0.008	0.002	0.000	0.000	0.000	0.000
	1	0.372	0.367	0.311	0.247	0.131	0.055	0.017	0.004	0.001	0.000	0.000
	2	0.124	0.275	0.311	0.318	0.261	0.164	0.077	0.025	0.012	0.004	0.000
	3	0.023	0.115	0.173	0.227	0.290	0.273	0.194	0.097	0.058	0.029	0.003
	4	0.003	0.029	0.058	0.097	0.194	0.273	0.290	0.227	0.173	0.115	0.023
	5	0.000	0.004	0.012	0.025	0.077	0.164	0.261	0.318	0.311	0.275	0.124
	6	0.000	0.000	0.001	0.004	0.017	0.055	0.131	0.247	0.311	0.367	0.372
	7	0.000	0.000	0.000	0.000	0.002	0.008	0.028	0.082	0.133	0.210	0.478

(continued)

Binomial probabilities:

$\binom{n}{x} p^x (1-p)^{n-x}$

Numbers in the table represent $p(X=x)$ for a binomial distribution with n trials and probability of success p.

		p										
n	**x**	**0.1**	**0.2**	**0.25**	**0.3**	**0.4**	**0.5**	**0.6**	**0.7**	**0.75**	**0.8**	**0.9**
8	0	0.430	0.168	0.100	0.058	0.017	0.004	0.001	0.000	0.000	0.000	0.000
	1	0.383	0.336	0.267	0.198	0.090	0.031	0.008	0.001	0.000	0.000	0.000
	2	0.149	0.294	0.311	0.296	0.209	0.109	0.041	0.010	0.004	0.001	0.000
	3	0.033	0.147	0.208	0.254	0.279	0.219	0.124	0.047	0.023	0.009	0.000
	4	0.005	0.046	0.087	0.136	0.232	0.273	0.232	0.136	0.087	0.046	0.005
	5	0.000	0.009	0.023	0.047	0.124	0.219	0.279	0.254	0.208	0.147	0.033
	6	0.000	0.001	0.004	0.010	0.041	0.109	0.209	0.296	0.311	0.294	0.149
	7	0.000	0.000	0.000	0.001	0.008	0.031	0.090	0.198	0.267	0.336	0.383
	8	0.000	0.000	0.000	0.000	0.001	0.004	0.017	0.058	0.100	0.168	0.430
9	0	0.387	0.134	0.075	0.040	0.010	0.002	0.000	0.000	0.000	0.000	0.000
	1	0.387	0.302	0.225	0.156	0.060	0.018	0.004	0.000	0.000	0.000	0.000
	2	0.172	0.302	0.300	0.267	0.161	0.070	0.021	0.004	0.001	0.000	0.000
	3	0.045	0.176	0.234	0.267	0.251	0.164	0.074	0.021	0.009	0.003	0.000
	4	0.007	0.066	0.117	0.172	0.251	0.246	0.167	0.074	0.039	0.017	0.001
	5	0.001	0.017	0.039	0.074	0.167	0.246	0.251	0.172	0.117	0.066	0.007
	6	0.000	0.003	0.009	0.021	0.074	0.164	0.251	0.267	0.234	0.176	0.045
	7	0.000	0.000	0.001	0.004	0.021	0.070	0.161	0.267	0.300	0.302	0.172
	8	0.000	0.000	0.000	0.000	0.004	0.018	0.060	0.156	0.225	0.302	0.387
	9	0.000	0.000	0.000	0.000	0.000	0.002	0.010	0,040	0.075	0.134	0.387
10	0	0.349	0.107	0.056	0.028	0.006	0.001	0.000	0.000	0.000	0.000	0.000
	1	0.387	0.268	0.188	0.121	0.040	0.010	0.002	0.000	0.000	0.000	0.000
	2	0.194	0.302	0.282	0.233	0.121	0.044	0.011	0.001	0.000	0.000	0.000
	3	0.057	0.201	0.250	0.267	0.215	0.117	0.042	0.009	0.003	0.001	0.000
	4	0.011	0.088	0.146	0.200	0.251	0.205	0.111	0.037	0.016	0.006	0.000
	5	0.001	0.026	0.058	0.103	0.201	0.246	0.201	0.103	0.058	0.026	0.001
	6	0.000	0.006	0.016	0.037	0.111	0.205	0.251	0.200	0.146	0.088	0.011
	7	0.000	0.001	0.003	0.009	0.042	0.117	0.215	0.267	0.250	0.201	0.057
	8	0.000	0.000	0.000	0.001	0.011	0.044	0.121	0.233	0.282	0.302	0.194
	9	0.000	0.000	0.000	0.000	0.002	0.010	0.040	0.121	0.188	0.268	0.387
	10	0.000	0.000	0.000	0.000	0.000	0.001	0.006	0.028	0.056	0.107	0.349
11	0	0.314	0.086	0.042	0.020	0.004	0.000	0.000	0.000	0.000	0.000	0.000
	1	0.384	0.236	0.155	0.093	0.027	0.005	0.001	0.000	0.000	0.000	0.000
	2	0.213	0.295	0.258	0.200	0.089	0.027	0.005	0.001	0.000	0.000	0.000
	3	0.071	0.221	0.258	0.257	0.177	0.081	0.023	0.004	0.001	0.000	0.000
	4	0.016	0.111	0.172	0.220	0.236	0.161	0.070	0.017	0.006	0.002	0.000
	5	0.002	0.039	0.080	0.132	0.221	0.226	0.147	0.057	0.027	0.010	0.000
	6	0.000	0.010	0.027	0.057	0.147	0.226	0.221	0.132	0.080	0.039	0.002
	7	0.000	0.002	0.006	0.017	0.070	0.161	0.236	0.220	0.172	0.111	0.016
	8	0.000	0.000	0.001	0.004	0.023	0.081	0.177	0.257	0.258	0.221	0.071
	9	0.000	0.000	0.000	0.001	0.005	0.027	0.089	0.200	0.258	0.295	0.213
	10	0.000	0.000	0.000	0.000	0.001	0.005	0.027	0.093	0.155	0.236	0.384
	11	0.000	0.000	0.000	0.000	0.000	0.000	0.004	0.020	0.042	0.086	0.314

Numbers in the table represent $p(X=x)$ for a binomial distribution with n trials and probability of success p.

Binomial probabilities:

$$\binom{n}{x} p^x (1-p)^{n-x}$$

n	x	p = 0.1	0.2	0.25	0.3	0.4	0.5	0.6	0.7	0.75	0.8	0.9
12	0	0.282	0.069	0.032	0.014	0.002	0.000	0.000	0.000	0.000	0.000	0.000
	1	0.377	0.206	0.127	0.071	0.017	0.003	0.000	0.000	0.000	0.000	0.000
	2	0.230	0.283	0.232	0.168	0.064	0.016	0.002	0.000	0.000	0.000	0.000
	3	0.085	0.236	0.258	0.240	0.142	0.054	0.012	0.001	0.000	0.000	0.000
	4	0.021	0.133	0.194	0.231	0.213	0.121	0.042	0.008	0.002	0.001	0.000
	5	0.004	0.053	0.103	0.158	0.227	0.193	0.101	0.029	0.011	0.003	0.000
	6	0.000	0.016	0.040	0.079	0.177	0.226	0.177	0.079	0.040	0.016	0.000
	7	0.000	0.003	0.011	0.029	0.101	0.193	0.227	0.158	0.103	0.053	0.004
	8	0.000	0.001	0.002	0.008	0.042	0.121	0.213	0.231	0.194	0.133	0.021
	9	0.000	0.000	0.000	0.001	0.012	0.054	0.142	0.240	0.258	0.236	0.085
	10	0.000	0.000	0.000	0.000	0.002	0.016	0.064	0.168	0.232	0.283	0.230
	11	0.000	0.000	0.000	0.000	0.000	0.003	0.017	0.071	0.127	0.206	0.377
	12	0.000	0.000	0.000	0.000	0.000	0.000	0.002	0.014	0.032	0.069	0.282
13	0	0.254	0.055	0.024	0.010	0.001	0.000	0.000	0.000	0.000	0.000	0.000
	1	0.367	0.179	0.103	0.054	0.011	0.002	0.000	0.000	0.000	0.000	0.000
	2	0.245	0.268	0.206	0.139	0.045	0.010	0.001	0.000	0.000	0.000	0.000
	3	0.100	0.246	0.252	0.218	0.111	0.035	0.006	0.001	0.000	0.000	0.000
	4	0.028	0.154	0.210	0.234	0.184	0.087	0.024	0.003	0.001	0.000	0.000
	5	0.006	0.069	0.126	0.180	0.221	0.157	0.066	0.014	0.005	0.001	0.000
	6	0.001	0.023	0.056	0.103	0.197	0.209	0.131	0.044	0.019	0.006	0.000
	7	0.000	0.006	0.019	0.044	0.131	0.209	0.197	0.103	0.056	0.023	0.001
	8	0.000	0.001	0.005	0.014	0.066	0.157	0.221	0.180	0.126	0.069	0.006
	9	0.000	0.000	0.001	0.003	0.024	0.087	0.184	0.234	0.210	0.154	0.028
	10	0.000	0.000	0.000	0.001	0.006	0.035	0.111	0.218	0.252	0.246	0.100
	11	0.000	0.000	0.000	0.000	0.001	0.010	0.045	0.139	0.206	0.268	0.245
	12	0.000	0.000	0.000	0.000	0.000	0.002	0.011	0.054	0.103	0.179	0.367
	13	0.000	0.000	0.000	0.000	0.000	0.000	0.001	0.010	0.024	0.055	0.254
14	0	0.229	0.044	0.018	0.007	0.001	0.000	0.000	0.000	0.000	0.000	0.000
	1	0.356	0.154	0.083	0.041	0.007	0.001	0.000	0.000	0.000	0.000	0.000
	2	0.257	0.250	0.180	0.113	0.032	0.006	0.001	0.000	0.000	0.000	0.000
	3	0.114	0.250	0.240	0.194	0.085	0.022	0.003	0.000	0.000	0.000	0.000
	4	0.035	0.172	0.220	0.229	0.155	0.061	0.014	0.001	0.000	0.000	0.000
	5	0.008	0.086	0.147	0.196	0.207	0.122	0.041	0.007	0.002	0.000	0.000
	6	0.001	0.032	0.073	0.126	0.207	0.183	0.092	0.023	0.008	0.002	0.000
	7	0.000	0.009	0.028	0.062	0.157	0.209	0.157	0.062	0.028	0.009	0.000
	8	0.000	0.002	0.008	0.023	0.092	0.183	0.207	0.126	0.073	0.032	0.001
	9	0.000	0.000	0.002	0.007	0.041	0.122	0.207	0.196	0.147	0.086	0.008
	10	0.000	0.000	0.000	0.001	0.014	0.061	0.155	0.229	0.220	0.172	0.035
	11	0.000	0.000	0.000	0.000	0.003	0.022	0.085	0.194	0.240	0.250	0.114
	12	0.000	0.000	0.000	0.000	0.001	0.006	0.032	0.113	0.180	0.250	0.257
	13	0.000	0.000	0.000	0.000	0.000	0.001	0.007	0.041	0.083	0.154	0.356
	14	0.000	0.000	0.000	0.000	0.000	0.000	0.001	0.007	0.018	0.044	0.229

(continued)

Numbers in the table represent $p(X=x)$ for a binomial distribution with n trials and probability of success p.

Binomial probabilities:

$$\binom{n}{x} p^x (1-p)^{n-x}$$

		p										
n	***x***	**0.1**	**0.2**	**0.25**	**0.3**	**0.4**	**0.5**	**0.6**	**0.7**	**0.75**	**0.8**	**0.9**
15	0	0.206	0.035	0.013	0.005	0.000	0.000	0.000	0.000	0.000	0.000	0.000
	1	0.343	0.132	0.067	0.031	0.005	0.000	0.000	0.000	0.000	0.000	0.000
	2	0.267	0.231	0.156	0.092	0.022	0.003	0.000	0.000	0.000	0.000	0.000
	3	0.129	0.250	0.225	0.170	0.063	0.014	0.002	0.000	0.000	0.000	0.000
	4	0.043	0.188	0.225	0.219	0.127	0.042	0.007	0.001	0.000	0.000	0.000
	5	0.010	0.103	0.165	0.206	0.186	0.092	0.024	0.003	0.001	0.000	0.000
	6	0.002	0.043	0.092	0.147	0.207	0.153	0.061	0.012	0.003	0.001	0.000
	7	0.000	0.014	0.039	0.081	0.177	0.196	0.118	0.035	0.013	0.003	0.000
	8	0.000	0.003	0.013	0.035	0.118	0.196	0.177	0.081	0.039	0.014	0.000
	9	0.000	0.001	0.003	0.012	0.061	0.153	0.207	0.147	0.092	0.043	0.002
	10	0.000	0.000	0.001	0.003	0.024	0.092	0.186	0.206	0.165	0.103	0.010
	11	0.000	0.000	0.000	0.001	0.007	0.042	0.127	0.219	0.225	0.188	0.043
	12	0.000	0.000	0.000	0.000	0.002	0.014	0.063	0.170	0.225	0.250	0.129
	13	0.000	0.000	0.000	0.000	0.000	0.003	0.022	0.092	0.156	0.231	0.267
	14	0.000	0.000	0.000	0.000	0.000	0.000	0.005	0.031	0.067	0.132	0.343
	15	0.000	0.000	0.000	0.000	0.000	0.000	0.000	0.005	0.013	0.035	0.206
20	0	0.122	0.012	0.003	0.001	0.000	0.000	0.000	0.000	0.000	0.000	0.000
	1	0.270	0.058	0.021	0.007	0.000	0.000	0.000	0.000	0.000	0.000	0.000
	2	0.285	0.137	0.067	0.028	0.003	0.000	0.000	0.000	0.000	0.000	0.000
	3	0.190	0.205	0.134	0.072	0.012	0.001	0.000	0.000	0.000	0.000	0.000
	4	0.090	0.218	0.190	0.130	0.035	0.005	0.000	0.000	0.000	0.000	0.000
	5	0.032	0.175	0.202	0.179	0.075	0.015	0.001	0.000	0.000	0.000	0.000
	6	0.009	0.109	0.169	0.192	0.124	0.037	0.005	0.000	0.000	0.000	0.000
	7	0.002	0.055	0.112	0.164	0.166	0.074	0.015	0.001	0.000	0.000	0.000
	8	0.000	0.022	0.061	0.114	0.180	0.120	0.035	0.004	0.001	0.000	0.000
	9	0.000	0.007	0.027	0.065	0.160	0.160	0.071	0.012	0.003	0.000	0.000
	10	0.000	0.002	0.010	0.031	0.117	0.176	0.117	0.031	0.010	0.002	0.000
	11	0.000	0.000	0.003	0.012	0.071	0.160	0.160	0.065	0.027	0.007	0.007
	12	0.000	0.000	0.001	0.004	0.035	0.120	0.180	0.114	0.061	0.022	0.000
	13	0.000	0.000	0.000	0.001	0.015	0.074	0.166	0.164	0.112	0.055	0.002
	14	0.000	0.000	0.000	0.000	0.005	0.037	0.124	0.192	0.169	0.109	0.009
	15	0.000	0.000	0.000	0.000	0.001	0.015	0.075	0.179	0.202	0.175	0.032
	16	0.000	0.000	0.000	0.000	0.000	0.005	0.035	0.130	0.190	0.218	0.090
	17	0.000	0.000	0.000	0.000	0.000	0.001	0.012	0.072	0.134	0.205	0.190
	18	0.000	0.000	0.000	0.000	0.000	0.000	0.003	0.028	0.067	0.137	0.285
	19	0.000	0.000	0.000	0.000	0.000	0.000	0.000	0.007	0.021	0.058	0.270
	20	0.000	0.000	0.000	0.000	0.000	0.000	0.000	0.001	0.003	0.012	0.122

Index

About the Author

Deborah Rumsey has a PhD in statistics from Ohio State University. Upon graduating, she joined the faculty in the Department of Statistics at Kansas State University, winning the distinguished Presidential Teaching Award. She eventually returned to Ohio State and is now an Associated Professor for the Department of Statistics. Dr. Rumsey is a Fellow of the American Statistical Association (ASA), and she has served on the ASA's Statistics Education Executive Committee. She's the author of *Statistics For Dummies* (Wiley), as well as other For Dummies books, and she has published many papers and given many professional presentations on the subject of statistics education. She is a founding member of CAUSE: The Consortium of Undergraduate Statistics Education, and the conference designer for USCOTS: The United States Conference on Teaching Statistics. Her particular research interests are curriculum materials development, teacher training and support, and immersive learning environments. Her passions, besides teaching, include her family, fishing, bird-watching, driving a tractor, raising baby farm animals, and Ohio State Buckeye football (not necessarily in that order).

Dedication

To my husband, Eric, and my son, Clint Eric — you are still my greatest teachers.

Author's Acknowledgments

I want to thank the people who made this workbook possible: Kathy Cox, for listening to my ideas for a workbook originally and for helping make these ideas become an even better reality; Lindsay Lefevere, for continuing to be a source of great support and help as my acquisitions editor for many books; Chrissy Guthrie, for her excellent help and for being a calm source of support as my project editor; Dr. Marjorie Bond, Monmouth College, for her thorough technical review and many helpful comments and suggestions; and Jennette ElNaggar, for her excellent copy editing. Thanks again to my friend Peg Steigerwald, who continues to cheer for me from the statistical sidelines, and to my friend Kit, for her constant support throughout the years. Thanks to my dad for showing me that hard work and determination will get you there. And to Mom: Thanks for showing me how to get there and still have a smile on my face.

Publisher's Acknowledgments

Executive Editor: Lindsay Sandman Lefevere

Editorial Project Manager and Development Editor: Christina Guthrie

Copy Editor: Jennette ElNaggar

Technical Editor: Dr. Marjorie Bond

Production Editor: Mohammed Zafar Ali

Cover Image: © Digital Genetics / Shutterstock

PERSONAL ENRICHMENT

9781119187790
USA $26.00
CAN $31.99
UK £19.99

9781119179030
USA $21.99
CAN $25.99
UK £16.99

9781119293354
USA $24.99
CAN $29.99
UK £17.99

9781119293347
USA $22.99
CAN $27.99
UK £16.99

9781119310068
USA $22.99
CAN $27.99
UK £16.99

9781119235606
USA $24.99
CAN $29.99
UK £17.99

9781119251163
USA $24.99
CAN $29.99
UK £17.99

9781119235491
USA $26.99
CAN $31.99
UK £19.99

9781119279952
USA $24.99
CAN $29.99
UK £17.99

9781119283133
USA $24.99
CAN $29.99
UK £17.99

9781119287117
USA $24.99
CAN $29.99
UK £16.99

9781119130246
USA $22.99
CAN $27.99
UK £16.99

PROFESSIONAL DEVELOPMENT

9781119311041
USA $24.99
CAN $29.99
UK £17.99

9781119255796
USA $39.99
CAN $47.99
UK £27.99

9781119293439
USA $26.99
CAN $31.99
UK £19.99

9781119281467
USA $26.99
CAN $31.99
UK £19.99

9781119280651
USA $29.99
CAN $35.99
UK £21.99

9781119251132
USA $24.99
CAN $29.99
UK £17.99

9781119310563
USA $34.00
CAN $41.99
UK £24.99

9781119181705
USA $29.99
CAN $35.99
UK £21.99

9781119263593
USA $26.99
CAN $31.99
UK £19.99

9781119257769
USA $29.99
CAN $35.99
UK £21.99

9781119293477
USA $26.99
CAN $31.99
UK £19.99

9781119265313
USA $24.99
CAN $29.99
UK £17.99

9781119239314
USA $29.99
CAN $35.99
UK £21.99

9781119293323
USA $29.99
CAN $35.99
UK £21.99

dummies.com

dummies®
A Wiley Brand